Theobald Smith, Veranus Alva Moore

Additional Investigations Concerning Infectious Swine

Diseases

Theobald Smith, Veranus Alva Moore

Additional Investigations Concerning Infectious Swine Diseases

ISBN/EAN: 9783337327798

Printed in Europe, USA, Canada, Australia, Japan

Cover: Foto ©berggeist007 / pixelio.de

More available books at **www.hansebooks.com**

U. S. DEPARTMENT OF AGRICULTURE.

BUREAU OF ANIMAL INDUSTRY.

BULLETIN No. 6.

ADDITIONAL INVESTIGATIONS

CONCERNING

INFECTIOUS SWINE DISEASES.

BY

THEOBALD SMITH, PH. B., M. D., AND VERANUS A. MOORE, B. S., M. D.

PUBLISHED BY AUTHORITY OF THE SECRETARY OF AGRICULTURE.

WASHINGTON:
GOVERNMENT PRINTING OFFICE.
1894.

U.S. DEPARTMENT OF AGRICULTURE.

BUREAU OF ANIMAL INDUSTRY.

BULLETIN No. 6.

ADDITIONAL INVESTIGATIONS

CONCERNING

INFECTIOUS SWINE DISEASES.

BY

THEOBALD SMITH, PH. B., M. D., AND VERANUS A. MOORE, B. S., M. D.

PUBLISHED BY AUTHORITY OF THE SECRETARY OF AGRICULTURE.

WASHINGTON:
GOVERNMENT PRINTING OFFICE.
1894.

LETTER OF TRANSMITTAL.

U. S. DEPARTMENT OF AGRICULTURE,
BUREAU OF ANIMAL INDUSTRY,
Washington, D. C., April 21, 1894.

SIR: I have the honor to transmit herewith the manuscript of the report which contains the results of some important investigations concerning infectious swine diseases. These investigations have been carried on for several years by my assistants in the laboratory and the present report contains results not heretofore published.

Very respectfully,

D. E. SALMON,
Chief.

Hon. J. STERLING MORTON,
Secretary.

LETTER OF SUBMITTAL.

U. S. Department of Agriculture,
Bureau of Animal Industry,
Washington, D. C., March 31, 1894.

Sir: I have the honor to submit herewith investigations dealing mainly with problems relating to hog cholera and swine plague. They have been carried on since 1889, and have had for their object the endeavor to find out upon what factors the great diversity in the characters of infectious swine diseases depends. The problems have presented themselves in the course of the more important investigations on swine diseases already published, and a solution seemed essential to any further progress in this work. To the report is appended a brief résumé of the results obtained, together with suggestions as to future lines of work which are most likely to yield valuable results.

Respectfully,

Theobald Smith,
Chief, Division of Animal Pathology.

Dr. D. E. Salmon,
Chief of Bureau of Animal Industry.

TABLE OF CONTENTS.

	Page.
The hog-cholera group of bacteria. By THEOBALD SMITH	9
Varieties of the hog-cholera bacillus	9
Other bacilli, not found in outbreaks of hog cholera, which belong to the same group	17
General characters of the hog-cholera group of bacteria	22
Two outbreaks of swine disease caused by bacillus choleræ suis ζ associated with the swine-plague bacillus.	27
Experiments on the production of immunity in rabbits and guinea-pigs with reference to hog-cholera and swine-plague bacteria. By THEOBALD SMITH and VERANUS A. MOORE	41
Hog cholera	42
The production of immunity in rabbits with cultures of attenuated hog-cholera bacilli	42
The production of immunity with sterilized bouillon cultures of hog-cholera bacteria	46
Experiments with sterilized agar cultures of hog-cholera bacteria	52
Experiments with the blood of hog-cholera rabbits sterilized by heat.	57
Experiments with the blood serum of rabbits made insusceptible to fatal doses of hog-cholera bacilli introduced under the skin	60
Does the blood of immune rabbits possess any bactericide or antitoxic power?	63
Experiment on guinea-pigs with blood serum from guinea-pigs immunized against hog-cholera bacteria	64
Swine plague	65
Experiments with sterilized bouillon cultures of swine-plague bacteria	65
Experiments with sterilized suspension of agar cultures of swine-plague bacteria	68
Experiments with the filtrate of agar suspensions	71
Experiments with defibrinated, sterilized blood of rabbits affected with swine plague	72
Guinea-pigs made insusceptible to swine-plague bacteria offer no resistance to hog-cholera bacteria and rice versa	74
Protective action of blood serum from immune rabbits	74
Does the blood serum of immune rabbits possess any bactericide or antitoxic power?	77
Conclusions	77
Hog cholera	78
Swine plague	79

Page.

On the variability of infectious diseases as illustrated by hog cholera and swine plague. By THEOBALD SMITH and VERANUS A. MOORE..... 81

Swine plague 82

Modifications of the septicæmia type by increasing the resistance of rabbits 83

Hog cholera 85

A form of pseudo-tuberculosis produced in rabbits and guinea-pigs as a result of increased resistance of the animals or attenuation of the virus 87

Can the bacillus of hog cholera be increased in virulence by passing it through a series of rabbits? By VERANUS A. MOORE 97

What becomes of hog-cholera and swine-plague bacteria injected in small numbers into the subcutaneous tissue of pigs? By VERANUS A. MOORE 103

Experiments with hog-cholera bacteria 103

Experiments with swine-plague bacteria 106

Practical bearing of the preceding investigations. By THEOBALD SMITH.... 109

ADDITIONAL INVESTIGATIONS CONCERNING INFECTIOUS SWINE DISEASES.

THE HOG-CHOLERA GROUP OF BACTERIA.

By THEOBALD SMITH.

Now that authorities, especially the German, have satisfied themselves that the bacteria described in former reports of the Bureau as hog-cholera and swine-plague are really distinct species belonging to different groups, and that the somewhat labored efforts to re-examine the work in this direction have come to a close, we may be permitted to examine this important group of bacteria a little more closely and report on certain variations of the hog-cholera bacillus itself as encountered in different outbreaks. The necessity for more extended studies of allied bacteria forming well-marked groups is painfully evident as we read over the now voluminous literature of bacteriology. Especially with regard to the apparently spontaneous appearance of epidemics and epizoötics of infectious diseases and the interrelation of human and animal diseases, it is of the utmost importance to examine, not only bacteria actually found to produce disease, but also those which are evidently closely related to disease-producing bacteria in their morphological and biological characters.

VARIETIES OF THE HOG-CHOLERA BACILLUS.

Bacillus choleræ suis.—This microörganism was first encountered in 1885 * and possessed the following characters, as determined at that time and subsequently:

Morphology.—A rod-shaped organism (bacillus) 1.2 to 1.5 μ long and 0.6 μ broad, with extremities rounded off. Spore formation not observed. It possesses the power

---

* Second Annual Report of the Bureau of Animal Industry (1885), p. 212. This organism was at that time called the swine-plague bacterium. The following year another organism was found in diseased swine so closely resembling the swine-plague bacterium, as described by Schütz in Germany a few months previous, that the name swine-plague was given to this second organism for the sake of uniformity of nomenclature and the old name "hog cholera" given to the bacillus described in this article. This change seems to have caused more or less misunderstanding and led to much misguided criticism.

of motion to an unusual degree. It stains readily in aniline dyes. In cover-glass preparations from the tissues of inoculated animals, the central portion of the rod is frequently only feebly stained.

Biology.—This bacillus grows fairly well on nearly all of the ordinary culture media, such as bouillon, gelatin, agar, milk, and blood serum. It thus possesses distinctly saprophytic properties. It is favored by the temperature of the blood and multiplies but feebly at 70° F. It fails to liquefy gelatin or to coagulate milk. It is destroyed in fluids at a temperature of 58° C. maintained ten to fifteen minutes.

Pathogenesis.—It produces in swine the infectious disease characterized chiefly by necrosis and ulceration of the mucous membrane of the large intestine and is occasionally accompanied by pneumonia of limited extent. It is fatal to rabbits, mice, and guinea-pigs in small doses, introduced either under the skin or into the stomach. In rabbits the disease is usually fatal in seven days when very minute doses are introduced under the skin. The period of disease may, however, be shortened *ad libitum* by employing large doses or injecting into the circulation directly.

Rabbits which have succumbed to minute doses show on post-mortem examination a small amount of suppurative infiltration at the place of inoculation. The spleen is engorged, the liver in a state of parenchymatous degeneration, and *contains small foci of necrosed tissue.* The kidneys are in a state of parenchymatous degeneration, the urine albuminous and containing hog-cholera bacilli. Peyer's patches may be pigmented and swollen. At the pyloric valve, the duodenum is frequently ecchymosed (probably due to discharge of bacilli in the bile). The heart muscle is markedly fatty, the lungs hyperæmic and frequently do not collapse when the thorax is opened.

More recently some additional characters have been determined which define still more narrowly its relation to other species and groups. Among these are the following:

1. The existence of a variable number of flagella. These have been specially studied in this laboratory by Dr. V. A. Moore.[*] He found most frequently 2 to 5, more rarely 8 or 9, flagella on a single organism. They vary in length, the more usual dimensions being from 7 to 12 μ. In one case a flagellum was found 18 μ long.

2. The fermentative action on dextrose in ordinary peptone bouillon. During the fermentation carbonic dioxide and hydrogen are evolved, and the fluid becomes markedly acid. Milk sugar and cane sugar are not attacked. Neither gases nor acids are formed in bouillon containing them.

3. A peculiar action on milk by which this fluid becomes opalescent and partly translucent.

We shall denominate the bacillus first described and subsequently encountered most frequently in outbreaks of swine disease as bacillus α and apply to the other varieties found other letters.

B. chol. suis β.—This variety was obtained from the spleen of a Nebraska pig early in 1886. It has been described in detail [†] in former publications and only a summary of its divergent characters is called for. One of the first things to arouse my interest was the bouillon culture. Within twenty-four to thirty-six hours after inoculation the fluid was covered by a thin, brittle, but complete membrane. The microscopic examination of such a culture showed that the bacilli in the fluid were of the same size as those of bacillus α and equally active.

---

[*] The Wilder Quarter Century Book, p. 355.

[†] Third Annual Report of the Bureau of Animal Industry (1886), p. 38.

In the membrane, however, the rods were evidently a trifle longer and not in motion.

There was another peculiarity observed which is of considerable interest. In making cultures on gelatin plates by drawing the platinum wire dipped in bouillon cultures rapidly across the still soft gelatin layer three or four times I observed that colonies of bacillus α appeared within due time, while those of bacillus β on the same plate refused to grow. With a low power of the microscope, very minute colonies could be detected, which grew no larger. The difficulty was found to be due to the fact that β requires a more strongly alkaline reaction of the gelatin than α. When this demand was satisfied the colonies grew even larger than those of α.

It is a curious fact that this variety, kept growing in the laboratory, mainly on agar, for over six years, still manifests these same peculiarities. It forms the membrane with the same rapidity, and unless the gelatin is distinctly alkaline* the colonies do not appear. A comparison made in December of 1892 with a number of cultures from various outbreaks showed that while colonies of all the other varieties of the hog-cholera bacillus appeared in gelatin rolls β refused to develop.

In order to make it grow sodium carbonate was added to the ordinary feebly alkaline beef-infusion-peptone gelatin and the result was quite striking, as the following notes show:

Na$_2$CO$_3$ in the form of a normal, sterilized solution was added to feebly alkaline nutrient gelatin in different quantities:

    a. To gelatin a, 0.025 per cent.
    b. To gelatin b, 0.05 per cent.
    c. To gelatin c, 0.075 per cent.

With these three sets of tubes rolls were prepared from the same bouillon culture. The first dilution was made in sterile water, the second and third in gelatin. The rolls made therefrom will be denominated *A* and *B* respectively. In *B* the colonies were few and became much larger subsequently than in the *A* roll.

4th day. *a. A* roll, colonies 0.032–0.048 mm. diameter.

    b. A roll, colonies 0.128–0.16 mm. diameter.

    c. A roll, colonies 0.128–0.16 mm. diameter.

13th day. *a. A* roll; colonies of all sizes, from mere specks to such as large as in *b* and *c.*

    B roll; only two deep colonies, each 1 mm. diameter.

    b. B roll; deep colonies 1–1.25 mm. diameter, surface colonies 3–3.5 mm. diameter.

    c. Same as *b.*

19th day. *a. B* roll shows more colonies than before; surface colonies as large as those of *b* and *c.*

    In all *B* rolls the surface colonies have formed large expansions resembling those of *B. coli.*

46th day. *a.* In *B* roll surface colonies have spread out in the form of thin patches 10–15 mm. diameter.

    c. In *B* roll they are more restricted, 5 to 6 mm. diameter, and fleshier.

---

* The culture medium is made alkaline with sodium carbonate, the indicator being litmus paper. For beef-infusion-peptone gelatin about 40 cc. of the normal solution is added to each liter.

The addition of the alkaline carbonate had thus the marked effect of causing the development of the colonies of β to such an extent that they outstripped those of all hog-cholera colonies I have ever seen.   In fact, the size of the surface colonies made me suspicious of some contamination until the inoculation of a rabbit with a culture made from one of these colonies, and various other tests, showed that my suspicions were unfounded.

A further peculiarity of the growth of these colonies appeared in the gelatin rolls containing the smallest quantity of alkali.   The colonies did not all appear at once and continue of the same size, as is universal in plate and roll cultures, but they appeared at intervals, so that, after a time, a large number of colonies, greatly varying in size, appeared where only a few had started originally.   Evidently the insufficient quantity of alkali added was responsible for this, for it did not occur in the rolls to which more alkali had been added.   In these the appearance of the colonies was simultaneous, and they continued to be of uniform size.*

The pathogenic power of these bacilli, after artificial cultivation for six and a half years, was not yet destroyed, for a rabbit which received into an ear vein 0.12 cc. of a bouillon culture died in four days.   The motility of this bacillus and the fermentative power in glucose bouillon were still present.

The pathogenic power of β was slightly less originally than that of α. It was fatal to mice and rabbits, but two guinea-pigs resisted inoculation.   It was fatal to a pig when a large quantity of bouillon culture was fed it.†

B. chol. suis γ.—Isolated from the spleen of a pig in 1890.‡   This variety is exceedingly interesting from the fact that its pathogenic properties are no greater than those of *Bacillus coli communis*.   That is, it requires fully 1 cc. of a bouillon culture, injected into an ear vein, to destroy a rabbit.   It is introduced here for the reason that in morphological and biological characters it differs but slightly from α.

Morphology.—In young bouillon cultures the bacilli appear slightly larger than α, and sluggishly moving chains of several elements are common.   In stained preparations the bacilli are slightly longer than those of α, measuring 1.8 to 2.5 μ.   Even filaments 4 to 8 μ in length were occasionally detected after the bacilli had been cultivated several years.

Biology.—The gelatin colonies are of the very delicate, spreading type, with a marked bluish cast in transmitted light.   They do not attain a greater diameter than 2 mm., and resemble closely colonies of *Bacillus typhi abdominalis*.   The bouillon culture becomes very turbid and has a faint odor.

Pathogenesis.—Feeble or absent.

* Colonies varying in size are also observed on plates made from cultures which have been subjected to the action of acids arising from fermentation.

† Loc. cit., p. 43.

‡ Report on Swine plague (1891), p. 77.

B. chol. suis δ.—Isolated at the same time with bacillus *γ*.

Morphology.—In general like *α*, though the bacilli appear a trifle plumper.

Biology.—Bouillon cultures become quite turbid. In the earliest cultures the bacilli formed clumps which in the hanging drop moved about as single masses. This peculiarity maintained itself for months and after passage through several rabbits. At present it seems to have disappeared.

Pathogenesis.—Quite feeble, perhaps a degree below that of *β*. The bacillus seems to have no effect on pigs.*

B. chol. suis ε.—This variety, obtained from a Virginia outbreak in 1890, agrees entirely with *α* in all but two features:

The surface colonies on gelatin plates instead of having the slightly convex form of *α* spread out in somewhat thinner, more or less bluish patches with sharp, undulating border, in this regard approaching the nonpathogenic variety, bacillus *γ*. A second differential feature is the viscid condition of the colonies and of the diffuse growth on agar. When this is touched with the platinum needle a delicate thread may be drawn out several millimeters in length. This peculiarity at first led me to suspect the purity of the culture. After making repeated plate cultures and passing the culture obtained from colonies through rabbits the same viscidity reappeared. It has persisted in cultures up to the present.

B. chol. suis ζ.—This interesting variety of the hog-cholera bacillus was obtained from cases of disease early in 1889. The outbreak presented a number of characters not noted in outbreaks examined up to that date. These have been very briefly referred to in former publications.†

Morphology.—This variety differs from *α* in appearing a trifle larger. In attempting, however, to present this impression of the eye in figures, I found it impossible to determine any difference, probably owing to the still crude state of the present microscopic measuring apparatus. It is not uncommon to find longer filaments in the various culture media, some of which show active movements.

Biology.—This variety may in general be regarded as nearer the saprophytic stage, for it grows more vigorously than *α* in the various culture media.

In gelatin plates and rolls the colonies appear within forty-eight hours, and at the end of a week develop into circular, smooth, glistening discs, which are grayish white and exhibit at times faint, concentric markings. When 1 or 2 cm. apart, they may be 3 or 4 mm. in diameter. The deep colonies are brownish, finely stippled spheres and attain a diameter of 0.4 to 0.5 mm. when not crowded. They thus resemble colonies of *α* in form and appearance, but are much larger.

The inclined agar growth differs from that of *α* only with reference to several minor points. The growth on the surface is the same with both germs, while in the condensation water there is a more active multiplication of the second germ, and the water assumes a milky white appearance. In cultures of *α* the liquid is not milky white, as the multiplication is far less active. In alkaline bouillon and bouillon with peptone this germ develops more rapidly and abundantly than does *α*. The liquid becomes very turbid over night in the thermostat and on shaking it a partly granular partly flocculent deposit arises. Careful inspection shows in some cultures a very delicate, iridescent pellicle on the surface. Multiplication was also observed

---

* Bulletin on Swine Plague, p. 80.
† Report of the Secretary of Agriculture for 1889, pp. 75–79.

in hay infusion. This was most vigorous when the hay infusion was made slightly alkaline.

On potato the growth resembles that of *α* very closely; it is perhaps of a more decided pale yellow than the latter. This description applies to cultures on pieces of potato in tubes from which evaporation is excluded. On some kinds of potato the growth is very feeble, whitish, and glistening, and easily overlooked. Bacillus *ζ* seems to be less sensitive to the reaction of bouillon. An abundant growth was obtained by adding to 10 cc. of the ordinary acid bouillon (not neutralized) a drop of lactic acid; *α* under the same conditions multiplies very feebly. This indifference to acid media may account for the fact that in the outbreak from which this variety was obtained the stomach was frequently attacked by necrotic and diphtheritic changes.

Pathogenesis.—This bacillus differs quite markedly from *α* in its virulence. The difference is best demonstrated by the inoculation of rabbits. A quantity of bouillon, less than one-fourth cc., did not prove fatal when injected subcutaneously, and in some cases even larger quantities were required. An early experiment with *α* showed that a dilution equivalent to $\frac{1}{100000}$ cc. of a peptone bouillon culture injected subcutaneously is still capable of proving fatal to rabbits.

While this comparatively large dose was required when injected subcutaneously, it was found in subsequent vaccination experiments with this variety that an equivalent of 0.02–0.01 cc. proved fatal when injected into an ear vein. The general character of the disease as revealed at the autopsy was likewise different from that produced by *α*. It lasted nearly a week longer. The spleen was not enlarged; the necrotic foci were not found in the liver; there was no hemorrhage in the duodenum, but, on the other hand, a striking disease of the intestines was present. The Peyer's patches of the small intestine were very much thickened and appeared as aggregations of whitish dots. The mucous surface over these patches was not infrequently covered by a slough. In the appendix vermiformis, part or all of the solitary follicles were enlarged, whitish, nodular, occasionally ulcerated. The Peyer's patches at the ileo-cæcal valve—one in the ileum, the other in the cæcum—were, as a rule, much thickened and covered by sloughs. In several cases the mucosa of the cæcum was covered with ulcers, probably due to bacilli discharged from the ulcerating Peyer's patches and localized here. The bacilli were readily demonstrated in the form of clumps in the infiltrated Peyer's patches and in some of the internal organs. The disease might be denominated typhoid fever of rabbits.

These various lesions did not always appear together in the same animal. Sometimes the ulceration was extensive, sometimes there was only swelling and infiltration of the follicles and Peyer's patches. Sometimes the disease was unusually prolonged. In one case, in which an ear vein injection had partially failed, the animal lived thirty-two days. All the solitary and agminated follicles of the intestines were found enlarged and infiltrated. The mucosa covering the patch in the ileum at the valve was necrosed.

In all cases there was a continuous elevation of temperature, the height of which varied with the intensity of the disease process, as gauged by the dose injected.

Feeding rabbits proved negative. The following experiments show that this variety affected mice, guinea-pigs, and pigeons:

Guinea-pig.—June 1, 1889. Two guinea-pigs received subcutaneously one-fourth and three-eighths cc., respectively, of a turbid peptone bouillon culture. The first guinea-pig died June 7. The chief lesion consisted in a sanguinolent and suppurative infiltration of the inoculated thigh, extending over the scrotum upon the other thigh. In the spleen a moderate number of bacilli. The second guinea-pig died June 10. There was a large ruptured abscess in the right groin. In the abdomen Peyer's patches are inflamed and thickened, the mesenteric glands infiltrated.

Pigeon.—May 27. Four pigeons were inoculated by partly subcutaneous, partly intramuscular injections of peptone-bouillon culture liquid over one pectoral muscle. Two received one-half cc. each, two three-fourths cc. Of these but one, which had received one-half cc., died on the following day. On the inoculated side the major portion of the pectoral muscle appears pale and necrosed. The internal organs present nothing abnormal. Bacilli present in the blood. In the other pigeons superficial sloughs formed in the pectoral muscle.

Gray mouse.—May 17. Three mice inoculated, from the same culture used for the pigeons, by subcutaneous injection of one-twelfth, one-eighth, and three-sixteenths cc., respectively. The third mouse dead next morning. The only discoverable lesion was the distension of a portion of the small intestine with a dark red mass, probably blood. The first mouse dies in about twenty-four hours with similar lesion. The second mouse was found dead on the seventh day after inoculation. The spleen was considerably enlarged and contained many bacilli. In the small intestine yellowish liquid contents. Peyer's patches appear from serosa as aggregations of blood-red dots. Culture on agar from the spleen pure.

Feeding mice.—Two mice were starved for a day, then fed with bread crumbs upon which the condensation water from the bottom of an agar culture had been poured. Both mice were sick next day and one died about noon. The spleen was very much engorged. In the small intestine a blackish, semiliquid mass (hemorrhage). No bacilli could be seen in preparations from liver and spleen. The second mouse survived this feeding as well as a second feeding attempted two weeks later. Two mice were fed in the same way a week later. Next day both were found lying on their backs, feet extended, breathing slowly and with difficulty and opening their mouths as if gasping for breath. Brought near a hot register they speedily recovered and were quite active the remainder of the day. On the seventh day both sick again and one died on the eighth day. In the slightly enlarged spleen hog-cholera bacilli were present, as demonstrated by an agar culture.

Bacillus ζ produces in swine a true hog cholera. The chief lesions are in the large intestine and in the stomach. In the outbreak in which this bacillus was encountered there was also found extensive broncho-pneumonia associated with swine-plague bacteria and lung worms. In pigs penned with those obtained from the outbreak the disease assumed a more decidedly intestinal type, the pneumonia disappearing. As the subject is of considerable importance the notes on three pigs of the original herd and of three of those infected by them are given in an appendix to this article.

The disease induced by ζ is more prolonged, more chronic than that produced by the virulent type, α. About four weeks elapsed between infection and death. The lesions of the stomach consisted of necrotic and diphtheritic changes of the mucosa usually outside of the fundus. In the large intestine the destruction of the mucosa was very extensive. The process was partly necrotic, partly exudative in character. The mucosa in the cæcum and upper colon was converted into an amorphous, whitish layer with irregular surface which I compared in the preliminary note to cork lining in appearance and texture. In the language of pathologists it would be called branny, i. e., as if bran had been sprinkled over the surface in a dense layer. Farther down the necrosis was less diffused, more localized, and now and then replaced by a true exudate, easily removed from a surface apparently but slightly altered. Large, button-shaped, neoplastic growths were encountered in one case of the original diseased herd.

The reduced virulence of bacillus ζ as compared with that of bacillus α was furthermore demonstrated by the feeding of cultures to pigs. Only softening of the feces and marked debility with loss of appetite were noticed for several days after the feeding, with the exception of one case, which proved fatal.

May 14, 1889. Pig No. 198.—Received three days ago from a farm upon which no swine diseases have existed for a number of years. Starved since yesterday morning. To-day a pint of a 4 per cent solution of crystallized sodium carbonate given to it of which only a portion is consumed. One hour later it is drenched with 400 cc., or nearly 1 pint of a bouillon culture in which bacillus ζ (derived from a colony in a gelatine roll) has multiplied for four days. The pig is not fed until next day. On the second day diarrhea sets in and the feces have a grayish color. Small quantities of a viscid yellowish mucus containing whitish flakes, are vomited occasionally. This condition persisted until death ensued quite suddenly four days after the feeding.

The autopsy was made 36 hours after death. The animal in the meantime was kept in the refrigerator. Male pig about 3½ months old. No skin lesions observed. On back of tongue near tip a group of whitish spots as of beginning necrosis. Stomach contains about a pint of yellowish flaky fluid. Mucosa over one-half of fundus dark red, swollen, and covered by a paper-like layer of a dull color. Around the cardiac expansion of œsophagus the mucosa has a dull yellowish appearance as if superficially necrosed. In the small intestine the inflammation progresses in severity from above downwards. This inflammation is manifested by increasing hyperæmia and injection of minute vessels, and by a tumefaction of Peyer's patches. In the ileum the walls are thickened and the reddened mucosa is covered with a thin layer of a yellowish white pasty exudate containing large numbers of leucocytes. The large Peyer's patch near the valve is partly hemorrhagic. In the large intestine the mucosa is reddened in patches, the walls of the gut thickened. The follicles in the lower colon appear with a central white point. When compressed these points are shown to be openings from which a plug of creamy matter exudes.

The urine in the bladder contains a trace of albumen and a large deposit of crystals of calcic oxalate.

From the spleen, slightly augmented in size, cultures were made, but remained sterile. From the mesenteric glands gelatin roll cultures and a bouillon-peptone culture were made. In these the fed hog-cholera bacilli developed.

Taking into consideration the pathogenic action of this bacillus on swine and on the smaller animals experimented with, no one will, I think, hesitate to accept the conclusion that bacillus ζ is a true hog-cholera bacillus closely related to or even specifically identical with bacillus α and that further experimental investigations are not called for ·to fortify this conclusion.

Five years ago, when this organism was first isolated, the attitude of bacteriolgoists, especially of the Koch school, was somewhat reserved in acknowledging the possible existence of distinct varieties of the same pathogenic species. In view of this a series of experiments were carried out to determine whether rabbits could be protected from the disease induced by bacillus α after passing through the disease produced by minute nonfatal doses of bacillus β. These results are briefly sketched on p. 43. They may be summarized as follows:

1. When α was reduced in its virulence the rabbits inoculated with it frequently presented the same gross lesions produced by β.

2. When rabbits were made partially insusceptible by inoculation with attenuated cultures of α and then inoculated with a virulent culture of α the resulting disease resembled that produced by β.

3. Rabbits which survived two inoculations of β were protected from a fatal disease when inoculated with α.

B. chol. suis η.—This variety was isolated by Dr. Moore in 1891 from the organs of a pig. It has been fully described in Bulletin No. 3 of this Bureau, and the reader is referred to that publication for a detailed account of its characters. It closely resembles α in all but one particular. It is nonmotile. This variation, apparently of such great morphological significance, is scarcely reconcilable with the close relationship existing between η and α in all other respects. Possibly η is a modification of α and originated at the Experiment Station of the Bureau where the pig was kept. Between 1885 and 1890 a large number of swine succumbed to experimental hog cholera at the station and it is not improbable that a few individuals of α may have survived in the soil or the body of insusceptible pigs where they gradually lost their motility. We know that motile bacteria may lose their motility wholly or in part in artificial cultures.

OTHER BACILLI, NOT FOUND IN OUTBREAKS OF HOG CHOLERA, WHICH
BELONG TO THE SAME GROUP.

Bacillus found in a mare after abortion.*—This organism must be
grouped with the true hog-cholera bacilli. The only recognizable differences were a tendency of the colonies in the earliest cultures to flow partly
down the inclined agar surface, and a feebler pathogenic power than
that of *Bacillus chol. suis α.*

It is obvious that these minor variations are not of sufficient importance to separate this bacillus from the hog cholera group. †

Bacillus enteriditis, Gærtner.—This organism was found in 1888 by
Gærtner in Saxony under circumstances which are of sufficient importance to deserve a somewhat extended quotation: ‡

A cow, suffering with diarrhea and discharging mucus, was killed and pronounced
fit for food. The autopsy revealed nothing abnormal beyond reddening of the small
intestine. The flesh appeared normal and was free from odor.

A young, strong workman ate 800 grams (less than 2 pounds) of this meat raw, and
seasoned with pepper and salt. He became ill two hours later, with vomiting and
diarrhea, and died in about thirty-five hours. The autopsy showed inflammation of
the small intestine. It was reddened and distended with gas. The solitary and
agminated follicles were swollen, whitish, the mucosa in places infiltrated, whitish,
in others reddish; the vessels markedly injected. In the stomach the fundus was
hemorrhagic.

The meat had been put on sale May 11, and up to May 18 fifty-eight persons belonging to 25 families became ill. All of these with one exception had eaten of this
meat. This person, the mother of the deceased young man, may have accidentally
infected herself with his discharges.

Of the 57 persons affected 12 had eaten the meat raw, 10 had eaten fried and boiled
liver, 2 a dish made from the lungs, 29 had taken boiled meat and soup, and 3 soup
only.

All persons who had eaten raw meat became ill, while of those who had consumed
cooked meat and soup about 36 remained well. The severity of the attack was proportional to the quantity of raw meat eaten. Thus, 1¼ pounds caused death in thirty-
five hours, while one-fourth pound led to a disease lasting fourteen days. In case of
the cooked meat and soup no such relation was noticed. The disease produced was
either very slight or severe and protracted.

There was no noticeable difference observed with reference to sex and age. The
disease appeared in most cases within twenty-four-to thirty hours with nausea, vomiting and purging; the vomited matters were frequently bloody, the stools greenish
and mucous. These symptoms were followed by severe general disturbance, such as
loss of consciousness, fever, and rapid, weak pulse, associated with great prostra-

---

* Bulletin No. 3 of the Bureau of Animal Industry, p. 53.

† The tendency of the colonies to run is easily accounted for by an increased production of a viscid substance around the bacilli. When sufficiently insoluble it
becomes visible as a capsule in some species. In certain bacteria cultivated for a
long time on agar the tendency to produce this substance seems to become augmented and a culture originally not viscid may become so after one or more years
of cultivation on alkaline agar. In the variety under consideration this tendency of
the colonies to run down the agar surface was transformed later into a tendency
to form a slightly wrinkled membrane. At present even this feature is no longer
recognizable.

‡ Corresp. d. allg. ärztl. Vereins Thüringen, 1888.

tion. Recovery took place in from five days to four weeks, according to the severity of the attack. During convalescence the epidermis peeled off.

In the flesh and the organs of the diseased cow and in the spleen of the deceased person the same bacillus was found. In the description of this bacillus by Gærtner no characters are recorded which separate this organism from the hog-cholera group as extended by the varieties previously described, excepting perhaps an unusually coarse granulation of the colonies on gelatin. Unfortunately, no mention is made of the action of this bacillus on milk.

The tests of this bacillus on animals showed both toxic and pathogenic action, and demonstrated the varying susceptibility of different species to this virus. Dogs and cats remained well after consuming freely of this meat. Similarly a fowl and sparrow were not affected after eating cultures of this organism. Gray and white mice fed with cultures died in from one to three days. Rabbits died after subcutaneous inoculation with particles of infected cow's flesh in from eight to twenty days. Guinea-pigs similarly treated recovered. Cultures acted in the same way.

In the inoculated animals the small intestine was usually filled with a thin, greenish-yellow fluid, and in some rabbits a glassy mucus was discharged. After intraperitoneal injection, fibrinous exudates, chiefly on the liver, were the rule. Occasionally the pleura was affected like the peritoneum. Not infrequently hemorrhages appeared, especially under the serous membranes. The local lesion was followed by very firm infiltration of the skin and subcutis, gelatinous œdema, and suppuration.

Gærtner further showed that meat in which these bacteria had multiplied, and which had been subsequently boiled to sterilize it, was very toxic and when fed to guinea-pigs and mice caused death. The same was true of the bouillon made from such infected flesh. In rabbits the poison did not act when introduced into the stomach.

Karlinski [*] describes a case of meat-poisoning, which he refers to *B. enteriditis* as the cause. It is a custom in Herzegovina to keep on sale, under the name *suche mieso* (dried meat), large portions of the carcasses of sheep and goats dried in the sun. This meat, although exposed to dust, sun, and rain for months, and although it is tasteless and has a penetrating rancid odor, is a favored article of diet among the poorer classes.

In May, 1889, a healthy man ate, as a result of a wager, 400 grams (somewhat less than a pound) of this dried meat softened by soaking beforehand. Within two hours nausea, vomiting, and purging set in. The temperature rose, the pulse became rapid and feeble, and the abdomen painful. Later, clonic spasms of the upper extremities, cold perspiration, and slight dilatation of the pupils were noticed. The temperature remained high for five days, then recovery slowly took place. The epidermis peeled off on the neck and extremities.

Karlinski isolated *B. enteriditis* from the vomited masses and from the dejections. Intravenous injections of this bacillus in small doses into quite young goats and lambs resulted in general depression and diarrhœa, followed by death in five days. Karlinski furthermore isolated this organism from several pieces of dried meat, from the intestines of several human beings, and a goat. From these encounters he draws the conclusion that *B. enteriditis* is a widely distributed organism.[†]

Lubarsch [‡] describes a bacillus in connection with a fatal disease in a new born child, which he is inclined to consider identical with *B. enteriditis*. There is, how-

---

[*] Centralblatt f. Bakteriologie u. Parasitenkunde, VI (1889), S. 289.

[†] Karlinski's frequent encounter of this bacillus raises the suspicion that he may have, at least in some cases, confounded it with varieties of *B. coli communis*. Our own experience is that bacilli of the hog-cholera group are very rare and isolated with great difficulty when saprophytic forms are also present.

[‡] Arch. f. pathol. Anat., CXXIII (1891), § 70.

ever, one fact stated in the printed observations which militates against this position, namely, the coagulation of milk. The results of inoculations on smaller animals are not very precise, the doses being chosen too large. Thus, the intraperitoneal injection of several cubic centimeters of a bouillon culture of the widely prevalent *B. coli communis* is likely to prove fatal in twenty-four hours. I have found 5 cc. of a culture of *B. coli* even after sterilization at 60° C. injected into the abdominal cavity of guinea pigs prove fatal within twenty-four hours in some instances.

A culture of *B. enteriditis* from Král's laboratory in Prague, Austria, was studied by the writer in 1893. The results of the examination are briefly as follows :

In gelatin rolls the surface colonies (3–10 mm. apart) were about 1 mm. in diameter on the fourth day. They appeared as slightly convex, round masses, coarsely stippled when viewed under a low power. The deep colonies on the same plate were opaque, brownish spheres 0.25 mm. in diameter. Little change took place in the rolls subsequently. In the gelatin stick culture the needle track developed into a slight growth in three days, the surface expansion was feeble.

On inclined agar a grayish, glistening smooth surface growth and opaque whitish deposit in condensation water not at all characteristic or peculiar.

Peptone bouillon became very turbid in three days. Odor feebly sour.

Milk was not coagulated.

On potato a dry, brownish-yellow growth had appeared in three or four days.

The bacilli from agar cultures are quite small and short; those from bouillon vary more or less in length and thickness. From potato they are very short, almost oval in outline.

In bouillon they are actively motile during the first few days, later on many are without motion. From an agar and a potato culture four days old they were nonmotile. Their behavior towards sugars does not differ from that of the true hog-cholera bacilli.

These morphological and biological characters are practically identical with those of the hog-cholera bacillus. The inoculation of rabbits likewise confirms this relationship.

July 12, 1893. White female rabbit, weighing 2¼ pounds, receives into an ear vein about 0.12 cc. of a bouillon culture twenty-four hours old.

July 16. Rabbit found dead this morning. Spleen moderately enlarged, dark, somewhat softened. Coccidiosis of gall bladder. No necrosis in liver. Kidneys pale. In appendix of cæcum a few enlarged whitish follicles. Quite a number of follicles of the lymphatic patch in cæcum near valve enlarged, whitish. Bacteria in spleen demonstrated by the microscope and in cultures.

July 12, 1893. White female rabbit, weighing 2½ pounds, receives into an ear vein 0.24 cc. of the same culture.

July 14. Dies this morning. Spleen soft, dark, moderately engorged. Kidneys hyperæmic; pale striæ in base of pyramids. Lungs hyperæmic. In spleen pulp a small number of bacteria resembling the hog-cholera bacillus closely. Cultures made therefrom contain this injected organism only.

The lesions of the intestines in the first rabbit resemble those produced by *B. chol. suis β.*

Bacillus typhi murium.—In October of 1890 this organism was found by Prof. Lœffler, of Greifswald, as the cause of an epizoötic among white mice kept for experimental purposes in cages at the Hygienic Institute. For account of the use of this bacillus in the destruction of field pests see the Annual Report of the Department of Agriculture for 1893 (p. 155). The description which Lœffler gives of this bacillus corresponds so closely with one or the other of the varieties of the hog-cholera bacillus that a detailed repetition of his statement is not called for.   The only differential characters requiring mention are the following:

Milk is stated to become distinctly acid, but to remain unchanged in appearance.

Guinea-pigs were slightly more susceptible to subcutaneous inoculation than rabbits.   The latter reacted with a local abscess only.

Two pigs, 4 weeks old, were fed with large quantities of bouillon. One remained well, the other died in eight days of an intestinal catarrh.   Although Loeffler is inclined to regard this death as due to some other cause, our experience would incline to the belief that the feeding was responsible for it.   A culture of this bacillus from Král's laboratory, examined in 1893, presented the following features:

In gelatin rolls the colonies appeared in two days.   On the fourth day the surface colonies (one-third to 1 cm. apart) had expanded so as to be 3 to 4 mm. in diameter.   The outline of the expansion was roundish, the margin delicately notched. The expansion was of nearly uniform thickness without central knob, the color grayish, partly translucent.   In some colonies there were faint indications of concentric zones of slightly varying opacity.

The deep colonies were spheres, about 0.5 mm. diameter, brownish, without coarse markings.   In rolls containing many crowded colonies, a distinct peripheral zone appeared.

The gelatin stick culture developed a slight growth in the needle track and a thin limited surface expansion.

On inclined agar, the appearance is the same as for *B. enteriditis.*

Peptone bouillon became exceedingly turbid in three or four days.

Potato culture precisely as that of *B. enteriditis.*

The bacilli in bouillon appeared as plump rods with rounded ends, some being mere ovals.   From agar they appeared as small as hog-cholera bacilli.

The motion in bouillon twenty-four and seventy-two hours old was very active. All appeared to be in motion.   From agar about one-third were in motion.   From potato culture four days old only Brownian motion detected.

Milk was not coagulated.

The fermentative properties were identical with those of the hog-cholera bacilli.

The following experiments on animals were made:

July 15, 1893.   Female rabbit, weighing 6 pounds, received subcutaneously 0.3 cc. of a bouillon culture one day old.

July 18.   Temperature 104.6° F.

July 22.   Temperature 102.8°.   Small local abscess.   Another rabbit which had received only 0.12 cc. of the same culture reacted in precisely the same way.

July 10, 1893.   A rabbit received into an ear vein 0.12 cc. of a bouillon culture six hours old.

July 11.   Temperature 106.4° F.

July 12.   Rabbit very ill; lies flat on abdomen, with head extended on floor of cage.

July 14.   Dead this morning.   Spleen very dark, soft, moderately enlarged.   Liver and kidneys quite pale.   Lungs hyperæmic.   Heart contains thick blood.

In spleen and liver many bacilli, usually in pairs and with a feebly-stained interior like hog-cholera bacilli.

Two gray mice fed with bread soaked in bouillon cultures on three separate occasions showed no signs of illness.

This bacillus, nevertheless, possesses a considerable degree of pathogenic power. In some inoculation experiments on guinea-pigs, made recently by Dr. C. F. Dawson under my directions, it was shown that while the guinea-pig withstood rather large subcutaneous doses (0.3 cc. of a bouillon culture) it was very susceptible to minute intraabdominal doses. In a more recent comparative experiment I found 0.02 cc. of a bouillon culture of B. typhi murium fatal within twenty hours when injected into the abdomen. The same dose of virulent hog-cholera bacilli injected in the same way proved fatal in not less than four days.

Another bacillus* recently described by Laser as causing epizoötics among field mice seems to belong to this group and to be perhaps closely related to B. typhi murium. A culture of this organism was not accessible for examination and comparison. Laser mentions that his bacillus produces acids in litmus whey equal in quantity to that produced by Emmerich's bacill us. This fact would group this bacillus with B. coli, for the hog-cholera group does not produce any acid in aërobic cultures of lac tose bouillon recognizable with litmus paper. Strangely enough, no state ment is made concerning its behavior in milk. A further differential character consists in its taking the Gram stain.

This bacillus appears to be more promptly fatal than hog-cholera bacilli, for both a field and a white mouse died in forty-eight and thirty-six hours, respectively, after subcutaneous inoculation. Feeding destroys the same species in four to six days. Similarly some rabbits and guinea-pigs proved susceptible. Neither the degree of susceptibility of these animals nor the pathological changes found at the autopsy have been sufficiently elucidated to be available for comparison with the hog-cholera bacillus.

A possible member of this group may be the bacillus found by Loeffler in 1881 as the presumable cause of diphtheria in pigeons.†

The disease had destroyed a flock of about 20 pigeons and only one case, the last, came into Lœffler's hands. He describes the sick pigeon as sitting trembling, with ruffled feathers and half-opened beak. The corners of the mouth, the base of the tongue, and the palate were covered with yellowish deposits. The dung was semifluid. The animal took no food, but drank much water. Death occurring during the night, the autopsy next day showed in addition to the deposits in the mouth a yellowish membrane extending through the trachea into the bronchi. In the lungs there were several pneumonic foci; there was parenchymatous clouding of the liver, slight enlargement of the spleen, occasional hemorrhages under the capsule of the kidneys. The mucosa of the intestines was markedly reddened and sprinkled with hemorrhages.

---

* Hugo Laser, Ein neuer für Versuchsthiere pathogener Bacillus aus der Gruppe der Frettchen-Schweineseuche. Centralblatt f. Bakteriologie, xi (1892), S. 184, xiii (1893), S. 643.

† Arbeiten a. d. kaiserl. Gesundheitsamt, ii (1884), S. 482.

The bacilli were found in the false membranes and in groups in the liver. The description is very meager. The inoculation of white mice is in so far interesting as it led to a fatal disease closely resembling that produced by hog-cholera bacilli in gray mice. The period of disease lasted from four to eight days. The spleen became very large and the liver mottled with necrotic foci.

GENERAL CHARACTERS OF THE HOG-CHOLERA GROUP OF BACTERIA.

After having given in brief a general description of the various members of the group of hog-cholera bacilli, we will bring together in compact form the underlying characters which bind the members of this group together, and those minor differences by which they have been individually differentiated. It should be stated that the following statements are based on an extended comparative study of these varieties, cultivated at the same time under precisely the same conditions. The microscopic observations were made only on groups of cultures prepared in this way. Careful measurements of the growth of colonies on gelatin plates have also been made, but the results, although showing well-marked constant differences, were not thought of sufficient importance to be given *in extenso*.

Morphology.—In form they are closely related. They are short rods with ends rounded, not producing spores, and possessing the power of motility with one exception (η).

More closely scrutinized they exhibit, in cultures, a certain divergence in size, which it is difficult to define. Thus bacillus ζ is evidently a trifle larger than α. In the membrane formed by β on bouillon the bacilli are longer than those in the bouillon and longer than α. The nonpathogenic form γ is distinctly larger than α, and cultures contain filaments of considerable length. Again, the swine-pest bacillus and the abortion bacillus resemble α very closely morphologically. *B. typhi murium* appears a trifle larger and plumper than α, and *B. enteriditis* inclines to involution forms.

Not only is it difficult to draw a sharp line when cultures of the same age and in the same medium are examined, but it is well nigh impossible to do this when various media are used. The different varieties vary in size from one medium to another, and any differences gained by the study of cultures in the same medium may be effaced by an examination extended to several culture media. The age of the culture as expressed by the time which has elapsed since its isolation from cases of disease may and does induce certain modifications of form not readily described.

In stained cover-glass preparations from the organs of inoculated animals, more particularly the spleen of rabbits, they resemble each other very closely. They usually present a more deeply stained periphery and give the impression of a bacillus completed filled out by a feebly-stained spore. This character, which I first pointed out in 1885, when taken together with their appearance in pairs is of considerable specific value. It is not always noticed in the most virulent varieties.

All the varieties which have been studied are capable of spontaneous motion excepting bacillus η. There is, however, no valid reason beyond this for excluding the latter from this group. In fact, setting this aside, it resembles α as closely as any other variety does.

Biology.—All varieties described fail to liquefy gelatin and as a rule grow with much less vigor on gelatin than the related group of *B. coli communis*. There exist among them 3 types of growth of the surface colonies on gelatin:

1. That of the virulent form α, the variety ζ, and *B. enteriditis*, consists of slightly

convex, roundish patches, varying in size according to the variety. The margin is not wavy, but uniformly convex. The surface very finely stippled.

2. That of the virulent variety ε, and the nonpathogenic γ, consists of a very thin iridescent expansion having an irregularly wavy margin. The surface shows a few branching radial furrows. This type of colony resembles closely that of the typhoid bacillus.

3. The type of θ and B. *typhi murium*, representing larger and coarser expansions with wavy margin simulating closely the surface colonies of B. *coli communis*.

It will be observed that these varieties do not correspond to any special degrees of virulence. In general, however, it may be said that with one exception (γ) the vigor of the surface growth is greatest with the least virulent varieties.

In bouillon the most virulent varieties (α, ε, η) and the abortion bácillus produce merely a moderate clouding. The less virulent varieties including the nonpathogenic form γ produce a considerable turbidity.

The reaction depends entirely on the presence or absence of muscle glucose derived from the meat from which the bouillon is prepared. If this is present the reaction becomes acid; if absent the reaction remains alkaline. The acid reaction becomes alkaline after a variable length of time, provided the quantity of sugar in the bouillon does not exceed 0.5 per cent. This change is owing to a base produced during the growth of the bacilli which slowly neutralizes the acid formed by the fermentation of the muscle glucose. *

On potato the growth is the same among the members of this group. There may be slight variations in the depth of color of the growth due to variations in the potatoes used. Now and then potatoes are encountered on which no growth appears. The surface assumes a glistening appearance but multiplication perceptible to the eye does not take place.

In milk no precipitation or coagulation of the casein occurs with any of the forms studied. There may be observed, however, if cultures remain in the thermostat at least several weeks, a gradually developing opalescence of the milk. The milk may even assume a light brownish appearance. During this prolonged stay in the thermostat the volume of the culture shrinks by evaporation and the opalescent fluid becomes quite thick but not viscid. It is markedly alkaline. The addition of acetic acid still causes precipitation of the casein. The process appears to be a kind of saponification of the fat as a result of the increasing alkalinity of the culture with age. This action of the hog-cholera bacillus was first noticed by Bunzl-Federn † and subsequently observed independently by Dr. V. A. Moore ‡ in milk cultures of the nonmotile variety of B. *choleræ suis η*.

The fermentative characters of the members of this group are remarkably uniform. I have tested them all in bouillon containing dextrose, saccharose, and lactose, respectively. Their behavior in the fermentative tube may be briefly defined as follows:

In peptone bouillon containing 1 per cent dextrose, gas appears within twenty-four hours and continues to form for three or four days. During the first twenty-four hours from one-third to one-half of the entire amount is formed, and at the end of forty-eight hours fully five-sixths has collected. During the third and fourth day only insignificant amounts are formed. The gas formation thus begins promptly and subsides with equal promptness. The total quantity which collects in the closed branch of the fermentation tube is equivalent to nearly one-half the capacity of this branch. The multiplication of the bacilli is rather feeble and subsides promptly. The strongly acid reaction developed during the fermentation is responsible for the cessation of growth and the subsequent destruction of the bacteria.

The gas set free is made up of carbonic dioxide and an explosive residue, probably hydrogen. The ratio of CO_2 to the latter is approximately as 1 to 2. This ratio

---

* On the method of determining the presence or the absence of muscle glucose in bouillon see the Wilder Quarter Century book (1893), p. 197.

† Arch. f. Hygiene, XII (1891), p. 198.

‡ Bulletin No. 3, p. 53.

holds only for the tube used and is different when gases are not allowed to escape. The reaction of the fluid is strongly acid.

A table illustrating the formation of gas in dextrose bouillon is subjoined:

Bacilli belonging to the hog-cholera group cultivated at 98° F. in bouillon containing one-fourth per cent peptone, 2 per cent dextrose, and one-half per cent sodium chloride.*

Culture.	Quantity of gas at 98° F. set free after—					Total 70°–80° F.	CO_2.	Explosive remainder (H).	Reaction
	One day.	Two days.	Three days.	Four days.	Five days.				
a	*Per ct.*	*Per ct.*	*Per ct.*	*Per ct.*	*Per ct.*	*Per ct.*	*Per ct.*	*Per ct.*	
Nebraska, 1889	17.4	41.7	54.8	60.0	57.4	54.8	38.0	62.0	Acid.
District of Columbia,1889	24.0	48.0	54.4	56.0	54.4	50.4	30.0	70.0	Do.
β									
Nebraska, 1886	32.5	35.8	37.5	40.0	37.5	36.6	34.0	66.0	Do.
γ									
New Jersey. 1890	33.3	48.3	51.6	56.6	55.0	50.8	37.7	62.3	Do.
∂									
New Jersey. 1890	40.8	50.0	53.6	56.0	53.6	50.4	38.0	62.0	Do.
ε									
Virginia, 1890	34.7	51.3	55.6	58.2	57.4	53.9	37.0	63.0	Do.
ζ									
Maryland, 1889	43.4	48.6	50.4	52.1	50.4	46.9	37.0	63.0	Do.
Virginia, 1889	45.0	58.2	58.2	58.2	55.8	52.4	36.0	64.0	Do.
η									
Non-motile bacillus				33.0	37.5	36.6	26.0	74.0	Do.
Abortion in mare	12.1	50.0	58.2	61.7	59.1	54.7	36.5	63.5	Do.
Swine pest	32.8	43.4	45.0	45.0	45.0	41.8	33.3	66.6	Do.
Mouse typhus		46.4	48.0		50.0	50.0	35.0	65.0	Do.
B. enteriditis		48.8	52.0		48.8	48.8	29.5	70.5	Do.

* There is no noticeable difference between 1 and 2 per cent dextrose bouillon.

This table demonstrates the marked uniformity among the members of this group in their behavior towards dextrose bouillon. Bacillus *β* and *η* produce somewhat less than the normal quantity of gas. It should be borne in mind, however, that the culture of *β* was over six years old when this gas test was applied. As to *η* it is not so easy to explain the variation.

A further evidence of the close relationship of the members of this group is manifested in saccharose and lactose bouillon. If the bouillon is free from muscle glucose not a trace of gas appears and the reaction, as determined with litmus paper, remains unchanged, or becomes more strongly alkaline. This must be regarded as sufficient evidence that these sugars are not attacked. The absence of any power to ferment lactose thus differentiates this group sharply from the colon group, some of which act upon lactose and saccharose as well as dextrose, others upon lactose and dextrose only. In both groups the action on dextrose is the same.

In order to facilitate the comparison of a considerable number of cultures with reference to the formation of indol Kitasato's method was used. Tubes containing 10 cubic centimeters of a sterilized solution of 1 per cent peptone and 0.5 per cent common salt in distilled water were inoculated with the various members of the hog-cholera group, and placed at 37° C. for from one to three days. The growth in all tubes remained very feeble. The presence or absence of indol was determined by adding to each culture 1 cubic centimeter of a 0.02 per cent solution of potassium nitrite, freshly prepared, and about 3 small drops of concentrated, chemically-pure, sulphuric acid. As a guide a number of cultures of *B. coli communis* were treated in the same way. The color appeared slowly and was most distinct after sixteen to twenty hours. While the cultures of *B. coli* assumed a decidedly red appearance, a few members of the hog-cholera group became feebly pink, the color of the culture of γ being the most accentuated. After a few days a decidedly carmine red deposit appeared in the cultures of *B. coli communis.* The deposit formed by the members of the hog-cholera group was whitish or feebly pinkish. This test thus gives for the colon group a different color than for the hog cholera group. It does not discriminate distinctly between members of the latter group and varies in intensity from almost complete absence of color to a decided violet pink, in cultures of γ, the nonvirulent variety.

In general cultures of hog-cholera bacilli are free from all odor. Among the more saprophytic varieties (γ, ζ) I have detected, in bouillon cultures contained in tubes permitting but little interchange of air, a faint sour odor.

Pathogenesis.—There is considerable divergence in pathogenic power among the varieties of hog-cholera bacilli and the other members of this group. A comparison of the results obtained by inoculation into rabbits shows this very clearly. Thus Bac. α is invariably fatal in very small subcutaneous doses. The same is true of Bac. ε. Bac. β also was originally of nearly equal virulence. Bac. ζ was originally fatal in some cases when rather large doses, 0.2 to 0.3 cc., of a bouillon culture were injected subcutaneously.

The rest of the group are fatal only when injected into an ear vein. For this end a small quantity (0. 1 cc. or less of a bouillon culture) has been usually sufficient. γ has no more pathogenic power than *B. coli communis* under like conditions. The original virulence of *B. typhi murium, B. enteriditis* and the swine-pest bacillus had undoubtedly been greatly reduced before they came into my hands. They were, however, still fatal to rabbits in small intravenous doses.

Besides a divergence in the degree of pathogenic power as determined by the quantity of culture fluid required to destroy rabbits there is a divergence in the nature of the pathological processes set up which is not wholly explained by a mere difference of degree.

Thus bacillus ζ produces in rabbits suppurative changes in the lymphatic apparatus of the intestines and the mesenteric glands in contradistinction to the necrotic and hemorrhagic lesions of the more virulent α. If we attempt to explain this as a mere difference of degree we are confronted by the abortion bacillus which is much less virulent than ζ, but when injected into the circulation produced marked necrotic lesions, in this respect standing nearer α. There is no doubt that with a reduction of virulence the necrotic give way in part to the suppurative processes, but this is not wholly true and we must accept subtle varieties of virulence not easily described. A marked illustration of this is the production of a peculiar pseudo-tubercular affection of the peritoneum in rabbits by a virulent hog-cholera bacillus of the α type attenuated by being contaminated with *Proteus vulgaris.* * Simple loss of virulence through long cultivation on artificial media does not produce such a peculiar modification.

A number of rabbits which were examined several weeks to months after recovery from a nonfatal dose of many of the bacteria described above showed in the appendix vermiformis and more rarely in Peyer's patches at the ileo-cæcal valve a few infiltrated follicles which appeared as whitish dots. This localization of the bacilli of this group in these follicles is perhaps the most constant macroscopic change to be credited to them. It appears only when the disease, after subcutaneous or intravenous inoculation, has lasted longer than a week or ten days.

In the table below an attempt has been made to bring together the minor variations among the members of this group and to omit all common characters. This table naturally suffers from the defects inherent in all tables attempting to formulate subtle differences. This is particularly true of any estimate of pathogenic activity. For a more accurate conception of this the reader is referred to the autopsy notes on rabbits in this and former publications.

* See p. 88.

Variety or species.	Original source.	Morphology.	Motility.	Surface colonies on gelatine plates.	Bouillon cultures.	General character of growth.	Pathogenic power with reference to rabbits. Degree as expressed in numbers.	Character.
B. choleræ suis a ...	Swine	Involution forms usually absent.	Active.....	Small, round, slightly convex.	Moderately clouded.	Only slightly vigorous.	1	Necrotic foci in liver.
B. choleræ suis β...	Swine	Somewhat longer forms in bouillon membrane.	Active.....	Large; spreading like B. coli.	Surface membrane; growth turbid.	Vigorous; requires more alkali.	2	Necrotic foci in liver.
B. choleræ suis γ ...	Swine	Involution forms common; slightly larger than a.	Active.....	Small, delicate colonies, resembling B. typhi abdominalis.	Turbid	Vigorous	Very feeble.	Like B. coli.
B. choleræ suis δ ...	Swine	A trifle plumper than a.	Active.....	Like a..............	Turbid; bacilli in clumps in early cultures.	Fairly vigorous....	4	Same as a.
B. choleræ suis ε ..	Swine	Like a..............	Active.....	Like γ.............	Like a.............	Fairly vigorous; agar growth visible.	1	Same as a.
B. choleræ suis ζ ...	Swine	Slightly larger than a ..	Active.....	Like a, but larger.....	Turbid	Vigorous	3	No liver necrosis; suppurative changes in Peyer's patches, etc.
B. choleræ suis η ...	Swine	Like a..............	No motility	Same as a............	Clouded	Like a..............	1	More speedily fatal than a.
Bacillus of abortion in mares.	Vagina of mare.	Like a..............	Motile.....	Same as a or nearly so.	Clouded	Like a..............	4	Necrosis in liver and spleen.
Swine pest.........	(Denmark)..	Like a..............	Motile.....	Same as a or nearly so.	Clouded	Like a..............	..	Same as a.
B. typhi murium Loeffler.	White mice (1890).	Plumper than a.........	Motile.....	Spreading like β.....	Turbid	Vigorous	4	No necrosis in liver.
B. enteriditis Gærtner.	Cow	Rather slender forms; variable.	Motile.....	Like a..............	Turbid	Vigorous	4	No necrosis in liver.

If we attempt to sum up those characters which are to circumscribe the hog-cholera group of bacteria we are at once confronted by the scarcity of common characters, as shown in the table above. Pathogenesis, though of great importance from the standpoint of pathology, is probably the last character acquired and evidently the most variable and most readily lost. If we base the unity of this group on morphological and biological characters, we are likewise met by variations in size, absence of motility, variations in the appearance of the colonies. There are, however, certain underlying characters, as expressed by the behavior of these bacteria in bouillon containing dextrose, saccharose, and lactose, which I think will serve as a very important group character, differentiating such group sharply from the colon group. I would therefore suggest that for the present all bacteria whose size approximates that of this group, which do not liquefy gelatin and whose fermentative properties are the same as those described for this group, should be ranged under it. Future investigations into the biochemical characters of these varieties or subspecies may reveal other differential characters, but the time has not yet come when such laborious work will be undertaken on a sufficiently extensive scale to be of any service in differentiating varieties and subspecies.

A question of considerable importance to which I wish to refer very briefly in conclusion is the origin of these varieties. Are, for instance, all the varieties of hog-cholera bacilli derived from one virulent form, or have they adapted themselves from a diversity of originally related saprophytic forms, such as the colon bacilli living under the similar conditions on mucous membranes, to a parasitic existence in the organs of the living animal?

Neither view is negatived by the information thus far at hand, but the second seems the more probable to the writer. A thorough discussion of this problem, so important to epizoötology, is premature, and it is simply suggested to call attention to its important bearings.

TWO OUTBREAKS OF SWINE DISEASE CAUSED BY BACILLUS CHOLERÆ SUIS ⸴ ASSOCIATED WITH THE SWINE-PLAGUE BACILLUS.

The following investigations, made four years ago, are now published for the first time. Their interest centers in the fact that in both outbreaks a much less virulent variety of the hog-cholera bacillus than that usually met with was found. The disease was likewise different from the ordinary acute hog cholera in several respects. It will be noticed that in nearly all cases swine-plague bacteria were present in the diseased lungs and in the intestines. Both outbreaks are thus illustrations of a mixed infection. In these investigations the information concerning the outbreaks was gathered by Dr. Kilborne, and the writer is indebted to him for the care of the animals at the Experiment Station and the notes made during the course of the disease. The writer is responsible for the pathological and bacteriological work.

I.

The history of this outbreak, as given by the owner, points very clearly to the importation of the disease from without.

On December 15, 1888, Mr. P., living near Knowles, Md., had eight fine shoats about 3 months old, and on this day he purchased a cheap lot of pigs in the Washington markets. One of these died on the way home, two others died during the two following days, and within ten days seven had died. The last one of the new lot died, greatly emaciated, January 20, after a sickness of from three to four weeks. The original lot on the farm showed signs of disease early in January, and up to January 22 four had died. Of the remaining four two are quite sick and two apparently well. Among the symptoms noticed by the owner was a rapid falling away in flesh, while the appetite remained fairly good up to the time of death. There was a severe cough, coupled with a nasal discharge and considerable diarrhea. In the latter stages of the disease the skin of the limbs, belly, and ears became deeply reddened. The ears turned almost black and "lopped, like the ears of a dog." On the limbs and belly the skin became "scabby, like a person with smallpox."

January 22. One of the original lot (No. 1) which died last evening was examined on the place. It was very much emaciated, the skin of ears, limbs, and abdomen a deep purple, shading into black. On the limbs numerous purplish crusts or scabs, one-sixteenth to one-eighth inch in diameter. The spleen somewhat congested, the liver pale; interlobular connective tissue increased in quantity; when cut a sensation of grittiness is imparted to the hand. The kidneys are pale, the surface with a few petechiæ. Lymphatic glands (with exceptions to be mentioned) in general slightly enlarged, with cortex dark colored. The lungs, stomach, and intestines were taken to the laboratory for a more careful examination.

Both lungs, with the exception of the caudal four-fifths of the principal lobes, solid, and moderately larger than the collapsed normal lung would be. The hepatization appears to be a cellular (catarrhal) plugging of the smaller bronchi and alveoli manifesting itself by a uniform mottling of the lung surface with yellowish dots. The extent of this catarrhal filling up is shown by the greater or smaller size of the yellowish dots, which in the cephalic lobes in part coalesce, so as to give the lung tissue a uniformly yellowish-gray appearance. The entire solid portion feels granular and hard. Imbedded in the normal tissue of the right principal lobe is a mass of solid tissue, as large as a hen's egg, in a more advanced state of degeneration. On section a number of sharply but irregularly outlined masses of a pale grayish-yellow appearance, evidently necrotic, make their appearance. In the left principal lobe are two small solid masses one-half inch in diameter. The pleura is not inflamed excepting over the solid mass in the right principal lobe. Here it is one-sixteenth inch thick, opaque, roughened, and wrinkled. On opening the air tubes the smaller bronchi are found very much dilated by thick purulent contents. When the latter is removed cavities appear from the size of a hemp seed to that of a bean, which are due to distention (bronchiectasis). The larger bronchi and the trachea contain a considerable amount of a dirty, muco-purulent, partly foamy liquid, through which are disseminated small whitish flukes. The mucosa is dusky, its minute vessels injected. In the caudal portion of the large bronchi and adjoining branches, numerous lung worms. Cover-glass preparations from the lung tissue show immense numbers of bacteria of several kinds.

The stomach is slightly pigmented in the fundus, and here two small superficial ulcers are found. Beyond the region of the fundus the mucosa is beset with minute papules about one-sixteenth inch across, somewhat paler than the surrounding mucosa. The small intestine is apparently intact, excepting the Peyer's patch in the ileum near the valve, which is swollen, pigmented, and has its surface pitted so as to give it a finely honeycombed appearance.

The mucosa of the large intestine is in general very much pigmented and hence quite dark in color. Throughout caecum and colon it is studded with sloughs from one-sixteenth to three-sixteenths inch in diameter, there being from five to ten to a square inch of surface. They are round, slightly projecting, convex. The projecting mass is of a dirty yellowish color and easily scraped away, as a pulpy or friable mass, exposing either a slightly raised or depressed rough surface almost white compared with the surrounding tissue. This firm mass extends only to submucosa. The mesenteric glands are slightly enlarged, bluish red; on section a reddened line can be seen under the capsule surrounding the gland tissue.

On the ventral surface of the kidneys were about fifty dark bluish spots (ecchymoses). On section the cortex pale, the medulla darkened, the minute vessels of the mucosa in the pelvis injected.

Sections of a slough from the large intestine, hardened in alcohol, examined subsequently under the microscope, showed that that portion of the tissue included in the slough failed to retain the stain. It involved the mucous crypts down to the submucous tissue. These were necrosed, although their original outline could still be faintly seen. Adhering to the surface of these and projecting slightly above the neighboring living tissue was a layer of exudate, consisting of amorphous matter not stainable. Beneath the layer of crypts the submucosa was more or less thickened by cell infiltration; the muscular coat intact.

Bacteriological examination.—From the spleen of this pig four cultures on agar were made, and in all a large number of isolated colonies appeared on the following day. These resembled hog-cholera bacilli colonies very closely. They were from one to three millimeters in diameter, according as they were crowded or scattered; pale grayish in color. In form and motility the bacilli could not be differentiated from hog-cholera bacilli. Subsequent observations showed slight difference, however, in the mode of growth in various media. That they were slightly different from bacillus α was shown by the following experiment:

January 29. From a bouillon peptone culture of this bacillus one-eighth cubic centimeter was injected subcutaneously into a rabbit. During the following three days the temperature rose 4° F. and then gradually fell to normal on the tenth day. It was killed on the fifteenth day, when fully recovered, and an abscess found in the groin extending to the abdomen and thigh and containing a rather consistent yellowish-white pus. The internal organs, including peritoneal cavity, appeared normal. Four bouillon peptone tubes were inoculated from the spleen and liver, as follows: Two received each nearly one-half of the entire spleen of the rabbit and two equally large pieces of liver tissue. One spleen and one liver culture became clouded after several days and contained the motile bacillus originally injected into the rabbit, and this only. Two remaining tubes were sterile weeks later.

January 23. A bit of lung tissue in which the disease process appeared most recent was torn up in sterile bouillon and one-fourth cubic centimeter injected under the skin of a rabbit. Cover-glass preparations show immense numbers of bacteria of several different forms. The rabbit was found dead January 28. There was considerable suppurative thickening of the subcutis and fascia on the inoculated thigh with some ecchymosis. In the abdominal cavity a slight amount of stringy exudate on the coils of intestines and some ecchymosis on the caecum. In the blood, spleen, liver, and peritoneal exudate numerous cocci, some of which show a polar stain. Agar cultures from peritoneal exudate and blood were placed at 37° C. The blood culture on the following day contained only swine-plague bacteria, the peritoneal culture other forms also.

At the same time another rabbit was inoculated in the same way with lung tissue which was in a more advanced stage of degeneration. This rabbit died January 28. There was an extensive pasty thickening of the subcutis on both aspects of the inoculated thigh, although there was no accompanying peritonitis. From the blood a pure agar culture of swine-plague bacteria was obtained.

A piece of the ulcerated large intestine had been kept in a sterile bottle on ice since the autopsy. On January 25 a rabbit was inoculated subcutaneously by injecting one-fourth cubic centimeter of a suspension of the ulcerated tissue torn up in sterile water. This rabbit died in four days with extensive subcutaneous pasty infiltration over the inoculated thigh, and peritonitis manifested by punctiform ecchymoses under the serosa of cæcum, membranous exudate on the peritoneal covering of spleen, liver, and cæcum. These lesions are characteristic of swine plague. An agar culture from the blood developed an abundant pure growth of swine-plague bacteria, while a culture from the abdominal exudate contained both the swine-plague bacteria and hog-cholera bacilli.

On January 26 a second rabbit was inoculated in the same manner from another bit of ulcerated mucosa. It was found dead on the eighth day. The subcutis of thigh, abdomen, and thorax were very much thickened with a pasty grayish yellow infiltration and extensively discolored by blood extravasation. Peritonitis slight, perhaps caused by the very many cysts of a cysticercus attached to the omentum. Spleen but slightly enlarged. From the blood two cultures, an agar and a bouillon culture, were prepared. In both only the motile cholera bacilli appeared. The results may be tabulated as follows:

Pig No. 1.

Spleen	lung	intestinal slough
Bacillus ζ rabbit	rabbit rabbit	rabbit
rabbit swine plague	swine plague Bacillus ζ	Bacillus ζ and swine plague
Bacillus ζ.		

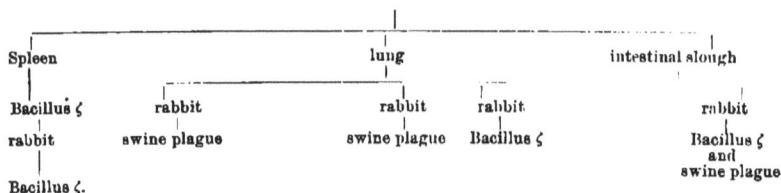

The two pigs from this herd which were diseased were taken to the Experiment Station January 22, and placed in carefully disinfected pens. One of these died on the following day.

Autopsy.—No. 2. White male, weighing about 60 pounds, in fairly good condition. A bright scarlet blush extending along both sides of the body as a band, 3-4 inches wide, and continued on the hind limbs nearly to the feet; the same blush on both ears and under surface of jaw and neck. Along the median line on the belly there is a strip purplish in color. The urinary meatus is stained by a subcutaneous hemorrhage. On the ventral aspect of the neck are 5 to 6 scabs one-fourth inch in diameter. These adhere very firmly, and covered by them are red, depressed spots, representing the true skin.

Superficial pubic glands about 2 inches long; lobules, pale red; very œdematous, the serum amber-colored as it flows from the cut surface. On opening abdomen, a large clot was found attached to the ventral abdominal wall near the spleen, and extending down between the coils of the small intestine; spleen very large, soft, dark. It is about 14 inches long, 3 inches wide, and three-fourths inch thick. The origin of the hemorrhage could not be traced. It may have come from the enlarged, congested spleen during the journey.

The stomach filled with food; mucosa normal; the small intestine normal. In the cæcum there are four neoplastic growths, one of them on the valve. There are six more in the upper 8 inches of the colon. The remainder of the large intestine has a normal mucosa, and is filled with lumps of hard, dry feces. The neoplasms, situated on a pale otherwise intact mucosa, are worthy of some attention. They are about three-fourths inch in diameter. Some project one-half inch above the mucosa, others are flatter, resembling a low, broad-rimmed hat. Cutting vertically through one of the projecting masses, we have first, a thin, coal-black surface layer, beneath this is a firm, palo yellow, almost leathery tissue, which forms the bulk of the projecting mass. The muscular coats beneath this are fused together, the neoplasm projecting somewhat deeper into them in the center than on the periphery. Lastly, the serous membrane under these masses thickened, discolored by old extravasations, and in some places adherent to adjacent structures.

The lungs are extensively diseased. Of the right lung the caudal tip and cephalic border of principal lobe, the ventral half of ventral and cephalic lobe are solid, and in the same condition as the lungs of pig No. 1. The surface mottling is faintly marked, owing to the pale, œdematous appearance of the diseased tissue. On the diaphragmatic surface of the principal lobe several areas of pneumonia are visible. Of the left lung the principal lobe contains a number of infiltrated lobules, the ventral lobe is entirely hepatized, its tip of a uniform grayish opaque color; the cephalic lobes contain a number of scattered hepatized lobules.

Trachea and bronchi almost occluded with molds of a pale yellow opaque mucopus. The terminal portion of both bronchi contain masses of lung worms.

In three agar tubes, inoculated by dropping into them bits of spleen tissue, no growth appeared. A rabbit inoculated with one-fourth cubic centimeter of a suspension of diseased lung tissue remained well.

The second diseased pig (No. 3), brought to the station January 22, was found dead on the morning of January 30. The lesions were limited to the lungs and large intestine, as the following notes show:

No. 3. Medium sized female, white. Decided reddening of skin over the pubic region. Of the right lung the cephalic and ventral lobe solid, granular to the touch. From the surface the lung tissue appears closely set with pale-yellow, minute spots. These give the tissue its granular consistency, and probably correspond to plugged and distended air vesicles. When a section is made through this solid tissue white consistent masses can be forced out of the air tubes, which are greatly enlarged in some places. In the cephalic lobe emphysematous lobules are interspersed in the solid tissue. In the principal lobe several foci of a grayish hepatization are present, while about one-half of the azygos lobe is solid.

Of the left lung the dependent half of the cephalic and the ventral lobe and about one-sixth of the principal lobe are hepatized. In the latter the disease is scattered in lobules having their bases chiefly in the diaphragmatic surface of the lobe.

Trachea and bronchi in part occluded with a very viscid, opaque, whitish mucopus mixed with foam, and containing fragments of lung worms.

Both sides of the heart contain large clots, dark in the center, pale outside where they touch endocardium, with branches extending into the large vessels.

Kidneys very pale. Liver gritty to the knife. The parenchyma softer than normal. Lymphatic glands in general swollen and slightly reddened.

The disease in the digestive tract is confined to the large intestine. The ileo-cæcal valve has its mucosa completely replaced by a layer of a homogeneous, cheese-like, rather firm material, whitish, the surface stained yellow. In the cæcum and succeeding 8 inches of colon are 8 roundish sloughs from one-fourth to one inch in diameter, of the same nature. Lower down in the colon are ulcers from one-eighth to one-fourth inch in diameter, about 6 to a square inch. Many of these are made up of an elevation or tumefaction of the mucosa about the size of a split pea, in the center of which is a depression filled with yellowish matter. Subsequent microscopic examination of these ulcers hardened in alcohol showed that they represented the solitary lymph follicles, swollen and ulcerated. Near the valve the large Peyer's patch presented the same appearance of ulcerated follicles in sections. The mucosa was everywhere of a dark, bluish-red color.

Bacteriological examination of the diseased lung tissue showed that the cheesy masses plugging the air tubes, which seemed to be made up exclusively of pus corpuscles, contained a large number of bacteria of different forms. A rabbit inoculated from some of the lung tissue torn up and suspended in sterile water did not become ill.

A rabbit which received an injection of a suspension, in sterile water, of a small portion of ulcerated tissue from the large intestine, died in three days from swine plague. At the place of inoculation the infiltration was slight. In the abdomen slight fibrinous exudation, hemorrhages under serosa of cæcum and rectum. Spleen congested. In the abdominal exudate, in the spleen and blood numerous bacteria, showing the polar strain well in cover-glass preparations from the blood. Agar cultures from spleen and blood contain an active vegetation of this germ on the following day.

The spleen of this pig, scarcely enlarged, showed no germs on cover-glass preparations. But three agar and one bouillon-peptone culture, made by adding bits of spleen tissue contained each on the following day an active growth of bacillus ζ. In two agar tubes the colonies were isolated in the upper part of the inclined layer, and from these cultures were made into other tubes.

From one of these second bouillon-peptone cultures a rabbit received into the thigh subcutaneously one-eighth cubic centimeter, February 1. The temperature of this rabbit on the following day was 104.2; on the 4th to the 8th, inclusive, it was 105.7, 106.8, 106.8, 104.5, and 102.6, respectively. The rabbit was killed two weeks after inoculation when evidently in good health. An abcess was found in the groin filled with partly curdy, partly putty-like pus. About one-third of the spleen dropped into each of the two bouillon peptone tubes. After several days both became clouded and contained only the motile bacillus injected into the rabbit. The following table summarizes the results:

Pig No. 3.

lung	spleen	intestinal slough
rabbit	Bacillus ζ	rabbit
(negative)		swine-plague bacteria

January 22. No. 121. A pig about three and one-half months old was placed in the pen with the preceding cases. One of these, it should be borne in mind, died January 23, the other January 30. Hence this animal came in contact with diseased animals as well as an infected pen. No. 121 was very much emaciated after three weeks' exposure; its respiration was labored and accompanied by a groan ; its back arched, flanks drawn in. The emaciation and weakness grew, and it died February 17, nearly one month after exposure. In the meantime several other fresh pigs had been added, and these attacked the carcass during the night and consumed part of the left lung and the heart. The autopsy made next morning revealed the following lesions:

Several hemorrhagic spots on abdomen; superficial sloughing of skin in patches on the sides of the thorax and abdomen. Also a number of scabs or crusts from one-eighth to one-quarter inch in diameter in the same situation, similar to those found on the other pigs. The spleen is barely changed in size and consistency. The lymphatics in abdomen small, slightly reddened, and pigmented. The serosa of large intestine of a more opaque white in patches and diffusely stained indicative of disease within the tube. Kidneys and liver apparently normal.

In the stomach the pyloric half of the mucosa of greater curvature is quite deeply congested. The small intestine is normal, the large intestine very severely diseased, however. Contents have the color and consistency of pea soup. The mucous layer of the entire cæcum and valve, with the exception of a small area in the blind end, is converted into a tough, yellowish-white, homogeneous layer, the free surface of which is very irregular. Its appearance might be compared to the roughness produced by cork lining, or by sprinkling small irregular fragments thickly over a surface. The entire thickness of the changed mucosa is about one-eighth inch, firmly bound to the muscular coat which appears as a dark red œdematous layer on section. The serosa is thickened and discolored.

Below the cæcum the necrosis breaks up into roundish ulcers gradually thinning out. There are about three to four to a square inch for the first 18 inches below ileo-cæcal valve. These ulcers are roundish, with a central plug of amorphous friable, orange-colored substance, very rough on the surface, surrounded by a border of a paler-yellow necrotic material. When the slough or exudate (whatever it may have been at the start) is scraped away, a white patch appears which represents the inflammatory infiltrate, partly necrosed, partly neoplastic, extending to the muscular coats. The greater part of the mucosa not destroyed by these sloughs is covered by thin, yellowish patches, not easily scraped away. When this is attempted the membrane beneath is pale and roughened. What remains of the membrane not covered by ulcerations and exudate is of a dark slate color.

The remaining 2 feet of colon has its mucosa entirely covered by a yellow, leathery, rather smooth diphtheritic membrane about one-sixteenth inch thick. When scraped away the membrane still adheres. The muscular coats are thickened, œdematous and redder than normal on section. The meso-colic glands are enlarged, the cortex pigmented.

The lungs are the seat of broncho-pneumonia. As already stated, the left ventral and cephalic lobe had been eaten by pigs in the same pen. The remaining principal lobe is œdematous, somewhat hypostatic. The dependent or free half of the right ventral and all of the cephalic lobe is solid, barely larger than in the collapsed normal condition. The pneumonia is identical with that found in the preceding cases, i. e., catarrhal plugging of air tubes, distention of alveoli by cell exudates, and hardening of the same, the peculiar surface-mottling with small, yellowish dots, being the result. The tip of the ventral lobe is covered with a thin sheet of grayish, fibrinous exudate. The cut ends of the small air tubes exude on pressure cylindrical curdy plugs, some of these showing on their surface the impress of the minute longitudinal folds of the mucous membrane. Through these two lobes are also disseminated about 12 necrotic foci, from one-eighth to one-half inch in diameter, consisting of greenish-gray homogeneous, rather firm caseous masses. The right principal lobe œdematous. The bronchi appear normal except in the diseased lobes. Here the membrane is bluish and, on pressure, the mouths of the branches exude a thick, yellowish fluid.

Microscopic examination.—Sections of the mucosa of the large intestine overlaid by the uniform pale-yellowish diphtheritic layer showed some very interesting characters. On the mucosa was an amorphous layer which failed to retain the stain (alkaline methylene blue) excepting along the free border where a line of blue was resolved into masses of saprophytic bacteria growing on the surface of the dead tissue. This exudative layer, about one-eighth millimeter thick, rested upon the mucosa proper, deprived of its epithelium, and at certain places the process of destruction had invaded the membrane as far as the fundus of the tubules, of which the outlines could barely be distinguished and which remained unstained. The submucous layer was considerably thickened by the infiltration of large numbers of leucocytes.

Careful examination under a high power showed that wherever the destructive process dipped down into the membrane itself, and in a section one-half inch long, this occurred five to six times; there a definite bacillus could be found. This bacillus pervaded the necrotic area and extended to the submucosa in large numbers. Wherever only the fundus of the tubules was intact, it was completely filled up by these bacilli forming large clumps of long, intertwining filaments. They were the only germs seen to extend into the normal tissue. Although appearing as long filaments, closer inspection revealed shorter forms, usually in pairs, each of which contained a paler central portion. The measurement of these long filaments diving

22275—No. 6——3

in and out of the tissue, and their resolution into segments was quite difficult. The short forms measured about 1.2 to 1.5 μ in length, and 0.5 to 0.6 μ in width.

Sections prepared from the deep circular sloughs show an unstainable necrotic mass extending almost to the muscular layers in the center and growing more and more shallow towards the periphery of the ulcer. Underlying the necrotic mass and extending a short distance from the slough is a zone of densely packed leucocytes causing a great thickening of the submucous tissue. In this ulcer neither the deeper layers of the slough nor the surrounding inflammatory zone contained bacteria.

Bacteriological examination.—From the spleen cultures on agar, in bouillon peptone, and in Esmarch tubes (gelatin) were made. In all the motile cholera bacillus ζ appeared, and in the bouillon tube also a streptococcus. The former must have been quite numerous in the spleen tissue, judging from the number of colonies in the roll cultures.

In order to test its pathogenic nature, a rabbit was inoculated February 20 by rubbing with a platinum loop some agar growth of this germ into the subcutis of abdomen. On the seventh day the temperature had risen to 105°, then it slowly fell back, reaching the normal on the fourteenth day. An abscess was forming meanwhile, which broke on the eighth day while taking the rabbit's temperature, and discharged pus freely. The rabbit was killed on the fifteenth day. Besides the abscess already noted, there was some enlargement of the spleen, but no germs were found in it on microscopic examination. The inoculated germ was still present in the spleen, however, for two bouillon peptone tubes, into each of which about one-third of the spleen was dropped, became clouded in forty-eight hours, with this and none other germ.

From a bit of diseased lung tissue four roll cultures are made in gelatin. The lung tissue had shown many bacteria not of one kind under the microscope. The first became completely liquefied; the others contained but a few colonies. One of these represents a micrococcus, another a slender motile bacillus. At the same time agar plates were made from the lung tissue. From these were isolated a large coccus, a streptococcus, and swine-plague bacteria.

At the same time a bit of diseased lung tissue was placed beneath the skin of a rabbit on the abdomen. It died in four days. There was extensive suppurative thickening and sanguinolent infiltration of subcutis of ab domen. Slight adhesion of cæcum to abdominal wall, with ecchymoses. Spleen small. In blood a few polar-stained bacteria found. In subcutis they were also seen amongst other forms. From culture of the blood and spleen several germs were obtained; one resembling swine plague, a second the hog-cholera bacillus, and a third nonmotile oval microbe. In order to identify the swine-plague germ in one of the impure agar tubes from the blood, a little of the swine-plague growth was placed under the skin of a second rabbit. Its temperature rose to 106.6° F. on the third day. On the seventh day it had fallen to 99.2° and it was found dead next morning. The lesions were those of swine plague. There was the local subcutaneous infiltration, whitish, pasty. The inoculation over the lowest ribs had caused extensive exudative pleuritis and pericarditis, the exudate being of a creamy consistency, covering the greater part of the lung tissue of right side, less extensive on left (inoculated) side. On the epicardium a similar deposit. Agar and bouillon peptone cultures from the spleen, blood, and pleural exudate made. All but that from the spleen fertile with swine-plague bacteria.

From the diphtheritic mucosa of the large intestine of the pig gelatin roll and plate cultures were made. That portion of the mucosa most recently diseased was taken, thoroughly washed by pouring sterile water over it and material from the scraped membrane taken. Of the gelatin roll cultures one became entirely the other partially liquefied. But there were a sufficient number of nonliquefying colonies intact to make cultures in different media. When developed they all contained the hog-cholera bacilli. On the plates there were very many nonliquefying among a few liquefying colonies. Subsequent examination with the microscope and in cultures showed that they represented hog-cholera bacteria. The following table summarizes the bacteriological examination:

Pig No. 121.

spleen	intestinal ulcer		lungs	
agar	streptococcus		gelatin rolls (negative)	agar plates
Bac. ζ	gelatin rolls			streptococcus swine-plague bacteria
rabbit	Bac. ζ			
Bac. ζ			rabbit	
	gelatin rolls	gelatin plates	swine-plague bacteria	Bac. ζ
	Bac. ζ	Bac. ζ		

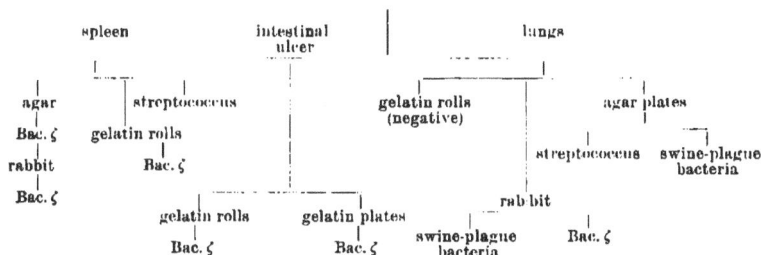

Two fresh pigs (Nos. 126 and 128) about four months old, in excellent condition, were placed in the same pen February 12, to keep up the disease for further study. No. 128 became emaciated within a week after the transfer and continued to grow feebler until March 5, when it was found dead. There were no distinctive symptoms excepting the wasting away.

Unfortunately the lungs had been entirely eaten out during the night by the other pigs in the pen, so that we have no knowledge of their condition. The spleen was but slightly swollen. In the liver there was stasis of bile in the larger ducts owing to plugging with ascarides which had forced their way up into the liver for some distance. The kidneys apparently normal.

The stomach is normal and filled with food; duodenum and jejunum, slightly congested, contain much yellow liquid. The mucosa of lower ileum, however, is very much reddened, and the smallest vessels visible to the naked eye distinctly injected. The gross lesions of the disease are limited to the large intestine. The mucosa is in a condition identical with that of No. 121. The entire cæcum involved. On section the mucosa is seen to be replaced by a dry, caseous, yellowish-white homogeneous layer, one-sixteenth inch thick, firmly bound to muscular coat. The free surface is very rough, yellowish, resembling the cork lining already mentioned. Below the ileo-cæcal valve about one-half of the membrane is involved, the caseous layer appearing in islands which are one-half to three-fourths inch square, easily scraped away as amorphous crumbling matter and leaving a decided depression. In the lower colon the deposit is less extensive and leaves scarcely any depression when scraped away. The mucosa where still intact presents here and there groups of petechiæ.

Bacteriological examination.—From the spleen of this pig cover-glass preparations were made, but the germs present were too few to be detected this way. Cultures on agar and in bouillon peptone from bits of spleen tissue contained next day bacillus ζ. To make sure, roll cultures in gelatin were made, and a rabbit inoculated from a culture derived from a colony. The illness of the rabbit was very slight. The temperature rose only to 103.8° F. on the third day, subsiding to 102.5° F. on the seventh. A small abscess had formed at the point of inoculation.

During the autopsy on the pig the bladder was ligated and removed with contents. For want of time it was kept on the ice for two days. Then the coats were burned through with a platinum spatula, several drops of clear urine added to gelatin and a roll culture made. After several days about 25 nonliquefying and several liquefying colonies appeared, the former made up of hog-cholera bacilli. Of the urine only about 10 cc. was found in the bladder. This was clear, pale-yellowish, acid, and containing much albumen. The deposit, after twenty-four hours, slight, somewhat viscid, consisting mainly of bladder epithelium.

No. 126, after being placed in the infected pen with No. 128, was found dead on the morning of March 6, i. c., in twenty-two days. The course of the disease was precisely the same as with No. 128.

Autopsy notes.—Small black female. Behind right shoulder the skin was destroyed over an area of 3 to 4 square inches, exposing the subcutis. On the skin over the right pectoral muscle a cluster of scabs about 1 inch square. The lungs are but slightly affected. A small portion of the free end of right cephalic lobe collapsed, bluish red. A small area of collapse in right ventral, another in right principal lobe. The lesions of the digestive tract closely resemble those of No. 128. The fundus of the stomach is considerably reddened. The duodenum contains several ascarides, one extending up the bile duct into the liver. The lower ileum considerably inflamed. The large intestine is extensively diseased. The mucosa is destroyed by a process identical to that observed in No. 128. Yellowish-white caseous transformation in large patches and bands in the cæcum and on the valve. Lower down in the colon exudative lesions still fresh; necrotic change of mucosa very slight. The rectum is involved as far as the anus in the same way.

Sections of the large intestine of pig No. 126 hardened in alcohol show about one-half of mucosa destroyed. Along a line half way between the base and the mouths of crypts is a densely stained line; beyond this only unstained amorphous débris attached to it, with free surface irregularly festooned. The same bacilli observed in sections of No. 121 are found dipping down in masses into the still living tissue, and wherever this dipping down occurs the tissue appears to be undergoing necrosis.

Bacteriological notes.—From the spleen of this pig the same hog-cholera bacillus was obtained pure in agar and bouillon peptone cultures, although too scarce to be detected on cover-glass preparations of spleen pulp. Its identity with the germs found in previous spleens was confirmed by its growth in gelatin roll cultures and inoculation into a rabbit. The temperature of the latter rose to 105.6° F. on the third day, but returned to 102° F. on the seventh day.

Agar plate cultures were made from a small bit of diseased intestine which was obtained as nearly as possible from the depths of the diseased membrane. From the plates the same motile germ was obtained, besides some liquefying forms. Plate cultures of gelatin were made at the same time. On these only a few liquefying colonies appeared. Of the others there were two forms, one growing on the surface as a very thin, spreading film, probably *Bac. coli communis;* the other appearing a day later and growing less vigorously, though by far the most numerous. This latter proved to be the hog-cholera bacillus. The presence of this germ was also demonstrated by the inoculation of two rabbits. A bit of the ulcerated intestine was placed beneath the skin of the thorax of each rabbit March 6. One died on the morning of March 12. The subcutaneous tissue of the whole abdomen and thorax was thickened and stained with blood, the skin was thickened and its under surface suffused with blood. The spleen slightly enlarged and congested. No peritonitis. The motile bacillus was seen in moderate numbers in cover-glass preparations and obtained in an agar culture from the spleen and the blood, about 25 to 50 colonies in each tube. A gelatine roll culture from the spleen contained a large number of one kind of colonies which were made up of the same germs. The second rabbit died next day, March 13. The lesions of the subcutaneous tissue were the same. The spleen was enlarged, dark, friable. Very slight ecchymosis and a few threads of fibrin on the cæcum. From agar and gelatin roll cultures of spleen and blood the same germ was obtained, but mixed with liquefying forms.

Pig No. 126.

spleen

| intestinal ulcer

Bacillus ς

agar plates   rabbit   gelatin plates

Bacillus ς   Bacillus ς   Bacillus ς

II.

During September of 1889 Dr. Kilborne was directed to examine reported outbreaks of swine disease in Clarke County, Va. The disease seems to have been lingering along the river below Berryville since the previous winter, and had become especially virulent during August. It was estimated that about 75 per cent of the swine in the vicinity of Berryville succumbed to the disease. At the time of the inspector's arrival there were very few sick, many having died during the two or three weeks preceding. The symptoms observed by the owners of herds were frequent coughing, diarrhea, and occasional vomiting. The affected animals lost flesh and strength very rapidly and affections of the skin were not uncommon. Pieces were said to drop off, leaving raw, deep sores. The skin of the ears and belly became purplish and was covered with sores.

Two pigs were killed by the inspector at the time which seemed diseased. The lesions in both were confined, as far as could be determined at the time, to a broncho-pneumonia, involving in one pig nearly one-half of the lung tissue. The lungs and spleens of both animals were brought to the laboratory. Culture from the spleen of one in bouillon peptone and on agar remained clear, nor could any bacteria be detected by microscopic examination of the spleen pulp. Cultures from the spleen of the second pig developed into a pure growth of a large bacillus with terminal spore which suggested malignant œdema. Two guinea-pigs inoculated from the culture remained well.

From the lungs of these two cases bits of diseased tissue placed under the skin of two rabbits without any result. On gelatin plate cultures from them only a few miscellaneous, liquefying colonies appeared.

Two diseased pigs were sent to the Bureau Station, so that more accurate observations could be made on the nature of the disease and its communicability to healthy pigs. These observations finally showed that the disease was caused by a hog-cholera bacillus identical with bacillus ς, and that whatever the communicable power of the disease had been it was very feeble, for a pig penned with the two cases of disease passed through a mild disease only. The post-mortem notes, bacteriological observations, and inoculation experiments which led to these inferences are as follows:

Pig No. 3, received September 4, 1889, and placed in disinfected, isolated pen. Animal at this time very thin and weak. Has been sick for a week according to statement of former owner.

September 6. Failing rapidly; diarrhea; passages dark, liquid.

September 7. Animal unable to rise.

September 11. Found dead this morning, and examined at once.

Autopsy notes.—Small female; very much emaciated. Over the sternal region the epidermis peels off in large patches. One ulcer, concave, one-fourth inch in diameter on the mucous surface of right side of upper lip, covered by a thin yellowish white slough; another smaller one on left side. On the tongue 3 similar ulcers, one in top and two on right and left border.

The mucosa of the fundus of stomach is intensely hyperæmic. The pyloric portion and upper duodenum deeply bile-stained. The entire length of the ileum is beset with depressed ulcers, of ragged, irregular outline, lined with a thin slough stained yellow. These ulcers are most numerous in the uppermost and lowest portion of the ileum. In caecum and upper foot of colon there are many small ulcers. They are present on the valve as well as on the Peyer's patch near it. In the lower colon they are few but large, often one-half inch in diameter. They are found in the rectum in small numbers.

Liver apparently normal. Bile in bladder thick, full of flaky material and containing several firm lumps of a dark brownish color. Kidneys with cortex pale, dull on section. The urine in bladder contains many short, finely granular casts, and a slight amount of albumen. The spleen is but slightly enlarged.

Lungs.—Both ventral and cephalic lobes, three-fourths of median lobe, one-half of right principal lobe, and a number of scattered foci of left principal lobe, hepatized. The color varies from a bright red to a pale red, mottled with faint yellowish dots. The hepatization is firm, granular to the touch. In trachea and bronchi abundant viscid muco-pus. Bronchial glands moderately enlarged, firm, of a pale pink color. Cover-glass preparations, from what appear to be the most recent diseased regions, show a large number of bacteria, resembling those of swine plague, within the protoplasm of the cells forming the exudate into the alveoli and minute air tubes.

The lymphatic glands near the stomach, in the angle of the lower jaw, those of the inguinal region beneath the skin, and under peritoneum near kidneys are all enlarged, reddened, mottled with paler spots, and with occasional petechiæ. The mesenteric glands have undergone the greatest change. They are all very much enlarged, of a peculiar yellowish green color. When the peritoneal covering is torn away, they appear made up of numerous whitish follicles.

Bacteriological observations.—No bacteria were detected in the internal organs with the microscope. Cultures on agar and gelatin plates were made from the liver, mesenteric gland, and lung tissue. Agar and bouillon peptone tubes were inoculated from the spleen, mesenteric, and inguinal glands. While the plate cultures indicated the presence of but very few germs in the various organs, the agar and bouillon cultures developed quite a variety. Each organ examined contained several forms. Thus in the bouillon peptone cultures from the blood a streptococcus and a very delicate true spirillum (not a comma bacillus) developed which could not be isolated. From the lung plates the swine-plague germ was obtained besides several other forms. In the spleen tubes the swine-plague germ appeared besides other forms. From the mesenteric glands *B. coli* was obtained. No pathogenic organism was thus found in this pig save the swine-plague germ, which was itself very much attenuated. One rabbit inoculated with a bit of diseased lung tissue died in nine days. There was extensive purulent infiltration of the subcutaneous tissue, starting from the place of inoculation and extending over abdomen and thorax. In the spleen and blood, the swine-plague germ only was present. In another rabbit, inoculated from a culture of this same germ, an abscess developed at the place of inoculation. This case was thus not satisfactory in spite of the large amount of time and labor spent upon it. Probably the hog-cholera bacteria had disappeared from the body, leaving it sufficiently weakened as a prey to other bacteria whose entrance into the closed cavities was favored by the numerous ulcers in the intestines. The following case is more to the point:

Pig No. 4 received at the station at the same time (September 4). It died late September 24, and was kept in refrigerator until next morning.

Autopsy notes.—Small female, weighs about 30 pounds. The mucosa of fundus of stomach intensely hyperæmic, associated with what appears to be slight, superficial necrosis. The entire length of the ileum has its mucosa covered with a layer of exudate, stained yellow, easily removed with the fingers. It is elastic, not readily broken up or torn, about 1 millimeter thick. In some places it adheres firmly enough to require scraping and pulling. The subjacent membrane is roughened, discolored. The walls of the intestine are much thickened. On serous aspect patches of punctiform hemorrhage. In the cæcum and colon there is superficial necrosis of the mucosa. Attached to it loosely, here and there, are small, shreddy masses of exudate, stained yellow. Only one ulcer found penetrating the mucosa. The walls of the large intestine considerably thickened and œdematous. Liver firmer than normal, with scattering lobules of a deep red color. On section, a gritty sensation imparted to knife. Bile in bladder very thick, owing to the suspension of scaly masses. Evident degeneration of cortex of kidneys. The urine free from casts and albumen. Spleen moderately enlarged. In it a considerable number of embolic foci, which feel like shot.

Both lungs symmetrically diseased. The ventral lobes, the median lobe, and about one-sixth of both principal lobes (median edge) hepatized, bright red, mottled regularly with faint yellowish points. In the cephalic lobes several small areas of collapse. From the small air tubes of the hepatized lobes protrude, on pressure, semisolid yellowish plugs of muco-pus. The larger tubes contain a more liquid secretion. In the main bronchi and trachea a small quantity of viscid muco-pus.

Bacteriological examination.—Cover-glass preparations of spleen pulp show a large number of rod-shaped bacteria, closely resembling hog-cholera bacilli in size and mode of staining. A bouillon peptone culture inoculated with spleen pulp is quite turbid on the following day and contains only motile bacilli like those found in cover-glass preparations. Two gelatin roll cultures made from a bit of spleen pulp showed within a few days an immense number of identical, nonliquefying colonies in the first tube and about 75 in the second. The latter spread on the surface of the gelatin as round disc-shaped bodies of a pearly luster reaching a diameter after one or two weeks of two to three millimeters. The gelatin was not liquefied at any time. The deep colonies are spherical, brownish by transmitted light with faintly granular disc from one half to three-fourths millimeter in diameter. In short, they appeared in all respects like Bacillus ζ.

In the liver, the same bacilli were found in large numbers in cover-glass preparations. A bouillon peptone and a gelatin roll culture contained the same bacilli found in spleen cultures and these only.

Gelatin roll cultures from the mesenteric glands contained colonies of the same bacilli and these only.

From the diseased lung tissue agar plates were made. These unfortunately dried out very rapidly after the first day at which date no growth was yet visible. At the same time a rabbit was inoculated by placing a bit of lung tissue under the skin. It died within thirty-six hours. The local lesion was slight and peritonitis absent. In heart's blood and spleen numerous swine-plague bacteria detected with the microscope. Cultures from these organs contain the same bacteria and these only.

Gelatin plates from diphtheritic patches of small intestine contained after a few days chiefly colonies of *B. coli.*

The hog-cholera bacilli were thus present in considerable numbers in the spleen, liver, and mesenteric glands. It was now necessary to determine whether this bacillus could produce disease, and, if so, the nature of the disease. For this purpose rabbits were used.

September 25. Rabbit received under the skin of the thigh, with a hypodermic syringe, one-quarter cubic centimeter of the bouillon culture derived from the pig's spleen. The temperature of the rabbit rose 2° to 3° F. during the week following inoculation, then gradually subsided. It was killed October 7, twelve days after inoculation, at that time apparently well. At the place of inoculation there was a small amount of purulent infiltration. The subcutaneous inoculation had thus failed, as it frequently did in case of Bacillus ζ from the Maryland disease. Intravenous inoculation was next tried with entire success, and any doubt as to the identity of this organism with Bacillus ζ thus vanished.

September 30. One-quarter cubic centimeter of a bouillon peptone culture derived from a colony in the spleen roll was injected into the ear vein of a rabbit. Its temperature on the following day 106.4° F. It was found dead October 3. The lesions consisted in a swelling of Peyer's patches and in the infiltration of the follicles in the cæcum, giving them the appearance of whitish dots. The spleen was engorged and contained the injected bacilli in considerable numbers.

From a bouillon culture obtained from this rabbit two fresh rabbits were inoculated, receiving into the ear vein one-eighth and one-sixteenth cubic centimeter, respectively. In order to gauge these small quanties accurately the culture liquid was diluted. The one which had received one-sixteenth cubic centimeter died in four days. The spleen engorged, patches of necrosis in the liver. Peyer's patches pigmented, rather prominent. Urine albuminous. The injected bacilli numerous in spleen and liver. Cultures confirmatory. The rabbit which had received one-eighth cubic centimeter was dead within three days. The spleen engorged, Peyer's patches, and follicles in the cæcum swollen. In the spleen numerous hog-cholera bacilli. Cultures confirmatory.

These small quantities still being sufficient to kill rabbits, one-fiftieth cubic centimeter bouillon culture derived from a spleen colony was injected into the ear vein of another rabbit October 12. It died October 17. A small number of follicles in cæcum are infiltrated, enlarged, and appear as whitish elevated dots from both serous and mucous surfaces. In Peyer's patches are a few follicles similarly affected. In the liver what appears to be necrosis in small patches. Spleen moderately engorged. In spleen and liver very few bacilli.

The contagiousness of this form of hog cholera was very feeble. On September 18, six days before the second diseased pig died, a healthy pig in good condition was placed in the same pen with it. The animal, apparently well for several months, became unthrifty. Its body was covered with thick crusts when it was killed, December 9. The only discoverable evidences of infection were several cicatrices in the large intestine and one healing ulcer to which a friable slough was still attached.

EXPERIMENTS ON THE PRODUCTION OF IMMUNITY IN RABBITS AND GUINEA-PIGS WITH REFERENCE TO HOG-CHOLERA AND SWINE-PLAGUE BACTERIA.

By THEOBALD SMITH and VERANUS A. MOORE.

In the following pages are brought together a series of experiments carried on during the past four or five years. They owe their origin to various causes, first among which was the necessity of learning something of the nature of the action of hog-cholera and swine-plague bacteria in the animal body. The failure to produce immunity in swine by the injection of small doses of virulent bacteria under the skin* made further studies of the problem of immunity necessary. This study could be best prosecuted on small animals, and for this reason rabbits and guinea-pigs were chosen.

Some of the investigations were made to examine into the claim of Metchnikoff† that immunity could be easily produced in rabbits towards hog cholera by the use of blood serum from immune rabbits. Our results had thus far been at variance with these, and proved to be totally so after repeating his work. Near the close of the investigation it was discovered that Metchnikoff had experimented with the swine-plague bacillus and not with the hog-cholera bacillus. This cleared up the difficulty. The detailed account of our observations will, without doubt, help to correct the wrong impression created by the misnaming of bacteria by Metchnikoff.

We have refrained from quoting or summarizing the now voluminous literature on the production of immunity in animals by the different processes which we have employed. The work is simply a contribution to the study of two specific pathogenic bacteria, and only those investigations are taken into consideration which deal with these bacteria. No claim is made for any new facts within the domain of immunity, excepting in so far as they apply to the disease germs experimented with.

The results obtained are, however, of special importance in shedding additional light on the divergent behavior of different species of bac-

---

* Report of the Bureau of Animal Industry for 1886, pp. 50-70.
† Études sur l'Immunité (5ᵉ mémoire); immunité des lapins vaccinés contre le microbe du hog-choléra. Annales de l'Institut Pasteur, VI, 1892, p. 289.

41

teria acting as causes of disease, as well as the marked difference in the receptivity of different species of animals towards the same disease germ.

HOG CHOLERA.

THE PRODUCTION OF IMMUNITY IN RABBITS WITH CULTURES OF ATTENUATED HOG-CHOLERA BACILLI.

Historically, the method of using attenuated cultures to produce immunity precedes that of other methods, and we, therefore, introduce the subject by detailing some experiments made in this direction.

These were begun as far back as 1889. The method of attenuation, based on that of Pasteur in the preparation of anthrax vaccines, consisted in exposing cultures in peptone bouillon to a temperature of 43.5° to 44° C. for a variable length of time. As the bacilli die out after a time in cultures at this high temperature, a fresh tube was inoculated every ten days, placed over night in a temperature favorable to multiplication, such as 37° C., and then restored to the temperature of 43.5° to 44° C. next day. This general plan suffered occasional interruptions, owing to disturbances in the thermostat employed, sudden changes of weather, which deranged the latter, and other accidents not infrequent in experimental work. From time to time rabbits were inoculated to observe any attenuation which may have taken place. During the process of attenuation certain changes in the appearance of the cultures were noticed. There was an increased deposit of bacilli which, when shaken up, appeared in flocculi and lumps of a whitish color. The great coherence of the bacilli, together with the tendency to form a brittle surface membrane, was not observed in cultures not exposed to heat. The bacilli appeared in small clumps in the hanging drop. Their motility was but slightly impaired and the clumps moved about very vigorously. After the systematic exposure to 43.5° to 44° C. had been stopped and the culture kept by inoculation from one agar tube to another, these modifications were still observed whenever fresh bouillon was inoculated. They had therefore become more or less permanent during the long-continued process of attenuation.

With the culture thus exposed to a high temperature rabbits were inoculated from time to time. After one hundred and seventeen days of exposure, comprising about eleven reinoculations into fresh bouillon, 0.1 cc. of the culture fluid was fatal to only one-third of the rabbits inoculated subcutaneously. The same may be said of the culture after one hundred and ninety-five days. In all rabbits there was noticed, however, a rise of temperature to 104° F. or 106° F. several days after inoculation. This fever continued a variable number of days.

Without wearying the reader with any detailed account of the animals treated we may briefly summarize the preliminary results by the statement that the immunizing action was not entirely uniform. Some treated cases died from inoculation with unattenuated cultures after a protracted illness; others fully recovered. The fact, however, was demonstrated that immunity in the very susceptible tame rabbit could be produced with cultures attenuated at a high dysgenetic tempera- ture. The process of inducing immunity proved to be a very tedious one and checkered with many failures.

The discovery in 1889 of a distinct variety of the hog-cholera bacillus (Bacillus ζ, p. 13) with less pathogenic power made it desirable to deter- mine whether rabbits could be made immune to the virulent variety after one or more inoculations with the less virulent variety. The opportunity for such a test was offered by the fact that Bacillus ζ was fatal to rabbits only when one-fourth of a cubic centimeter or more of a bouillon culture was injected subcutaneously. A preliminary attenua- tion was thus unnecessary. The method followed was to inject up to 0.2 cc. subcutaneously or from .003 cc. to .02 cc. (diluted) into an ear vein. When one or two months later the usual fatal dose of virulent hog-cholera bacilli was injected under the skin only 3 out of 8 rabbits survived. It is probable that if a second injection of the weaker variety had been made before the test inoculation the majority would have survived. This experiment thus confirmed the preceding, in establishing the fact that immunity could be produced in this way.

From these two series of experiments of 1889-'90 two rabbits were still living in 1892. They had been kept for further investigation, but other work interfered. During this interval they had successfully resisted the subcutaneous injection of the usually fatal dose of hog- cholera bacilli at two different times. The tags belonging to these rab- bits had been lost, so that it was impossible to determine to which series they belonged. Their continued immunity was demonstrated by the following inoculations:

July 29, 1892. Rabbit A., large white animal, weighing 5 pounds; subcutaneous inoculation of 0.1 cc. of a virulent bouillon culture.

August 1. 107° F.

August 4. 103.9° F. Swelling one-half inch thick at seat of inoculation.

August 13. 103° F.

September 15. 102.6° F. Subcutaneous inoculation with 0.2 cc. of a virulent cul- ture.

September 17. 103° F.

September 19. 103.2° F.

September 21. 103° F.

October 5. Intravenous injection of 0.06 cc. of a virulent bouillon culture.

October 7. 102.2° F.

This animal, supposed to have recovered entirely, died suddenly on November 8, over a month after the last inoculation. The autopsy was negative, excepting as to the brain. In the caudal portion of the right hemisphere there was a firm whitish tumor about 6 to 7 mm. in diameter. The microscopic examination of sections of hardened brain tissue revealed an abscess containing clumps of hog-cholera bacilli. The brain was examined because the person in charge of the animal room reported that the head of the rabbit had been drawn to one side one or more days before death.

July 29, 1892. Rabbit B, large yellow male; weighs 5.37 pounds. Subcutaneous injection of 0.1 cc. of a virulent bouillon culture.

August 1. 103° F.
August 4. 102.8° F.
August 13. 103° F.
September 15. Subcutaneous injection of 0.2 cc. of a virulent culture.
September 17. 103.4° F.
September 19. 102.6° F.
September 21. 102.8° F.

It is not improbable that the age of these two rabbits contributed, to a slight extent, to their continued powers of resistance.

From 1889 to 1892 no new work was attempted in this direction. In the latter year several things conspired to make a repetition of this work desirable. A remarkably weakened culture of the virulent hog-cholera bacillus attenuated in cultures, containing *Proteus vulgaris*, was in our possession. Metchnikoff had just published an article claiming to have readily produced immunity in rabbits by injecting the blood serum from rabbits previously immunized, by another method, against the hog-cholera bacillus. Improbable as these results seemed, they yet demanded reëxamination in the interest of some method of vaccination for swine.

Hence the attenuated hog-cholera culture was employed in the immunization of rabbits which were to furnish the blood serum to be used in repeating a portion of Metchnikoff's work.

The attempts at producing immunity were fraught with many failures, and only two rabbits were finally made insusceptible. The failures were due at first to a too rapid process of immunization, later on to a gradual return to virulence of the vaccinal culture, and to the accidents incidental to keeping animals for a long period of time. In some cases rabbits passed successfully through two inoculations, to be killed by the third, more severe test. The following is the history of the two rabbits which acquired a pretty high degree of resistance.* The details are of sufficient interest to warrant reproduction here.

White rabbit, No. 205, weight about 3 pounds at the beginning of experiment.

1. January 23, 1892. Receives a subcutaneous injection of 0.2 cc. of vaccinal culture; lesion at point of injection slight; temperature rose only to 103° F.
2. April 13. Subcutaneous injection of 0.15 cc. of same culture.
   April 18. 103.4° F.
   April 21. 103° F.
3. May 21. Intravenous injection of 0.04 cc. of same culture diluted in 10 volumes of sterile bouillon.
   May 23. 105.2° F.
   May 25. 102.8° F.
   May 27. 102.4° F.
4. June 30. Intravenous injection of 0.1 cc. of same culture.
   July 1. 106.8° F.
5. July 29. Subcutaneous injection of 0.1 cc. of a virulent bouillon culture of hog-cholera bacilli.
   August 1. 103.9° F. Seat of inoculation reddened and swollen.

---

* For the experiments to determine the protective power of the blood serum from these rabbits see p. 61.

August 4. 103.8° F. Swelling as before; control rabbit dead.
August 12. 103.6° F.

6. September 15. Temperature 103° F; subcutaneous injection of 0.15 cc. of a virulent bouillon culture of hog-cholera bacilli.*
September 17. 103° F.
September 19. 102.8° F.
September 21. 102.4° F.
October 20. Weight, 3⅞ pounds.
November 15. Weight, 4 pounds.

White male rabbit, No. 232, weighing 3 pounds.

1. February 6, 1892. Subcutaneous inoculation with 0.2 cc. of vaccinal bouillon culture.
February 8. 102.5° F.
February 11. 104° F.
February 13. 106° F.
February 15. 104° F.
February 17. 103.4° F.
February 18. 103.7° F.
February 23. 103.4° F.: firm swelling about one-half inch thick at seat of inoculation.
March 2. 102.4° F.

2. May 21. Intravenous injection of 0.04 cc. of vaccinal culture.
May 23. 100.4° F; rabbit apparently quite ill.
May 25. 102.4° F; rabbit better.
May 27. 103° F.

3. June 30, intravenous injection of 0.1 cc. vaccinal culture.
July 1. 104.2° F.

4. July 29, subcutaneous injection of 0.1 cc. of a virulent culture.
August 1. 103.7° F; red swelling at seat of inoculation.
August 4. 103.5° F; (control rabbit dead).
August 13. 103.4° F.
November 15. Weight 3½ pounds; has not gained during the process of vaccination.

These two rabbits were thus made insusceptible to a fatal subcutaneous dose of hog-cholera bacilli. No. 205 was twice tested with such a dose; No. 232 but once. The following case is interesting, for the vaccination was entirely intravenous, and the degree of immunity acquired was quite high.

No. 281, gray and white rabbit, weighing 3 pounds.

1. May 21, 1892. Intravenous injection of 0.02 cc. of the vaccinal bouillon culture, to which 10 volumes of sterile bouillon are added.
May 25. 105° F.
May 27, 104° F.

2. June 30. Intravenous injection of 0.06 cc. vaccinal culture.
July 1. 106.8° F.

3. July 29. Intravenous injection repeated; some of fluid passed into subcutis of ear; probable intravenous dose 0.1 cc.
July 30. 103.6° F.
July 31. 104.3° F.
August 4. 103.7° F.
August 12. 102.8° F.

* This stock culture being in constant use and found invariably fatal a control rabbit was not inoculated at this time.

4. September 15, weighs 4,³, pounds. Intravenous injection of 0.1 cc. of vaccinal culture.

September 17, 107.3° F.
September 19, 102.8° F.
September 21, 102° F.

5. October 5, subcutaneous injection of 6.06 cc. of a virulent bouillon culture diluted with sterile water.

October 19, rabbit dies to-day. Compression of spinal cord by ulcerated lumbar vertebræ. Either result of injury or of localization of injected bacteria. Paralysis of lower extremities was noticed some days before death. Even if we assume that the disease of the vertebræ was due to the injected bacteria, which is highly improbable, the injection having been made into the subcutis, there was still a decided resistance, since death occurred from ten to twelve days later than with unprotected rabbits.

THE PRODUCTION OF IMMUNITY WITH STERILIZED BOUILLON CULTURES OF HOG-CHOLERA BACTERIA.

The culture medium employed in these experiments, with one exception, was beef infusion, containing one-half per cent peptone, one-half per cent sodium chloride and a sufficient quantity of a normal solution of sodium carbonate to give it a faintly alkaline reaction as determined by litmus paper. It was sterilized by discontinuous boiling in tubes containing 30 cc. each. After their sterility was tested the tubes of bouillon were inoculated with hog-cholera bacteria and placed in an incubator at a temperature of 35–37° C. After a certain number of days (the cultures used varied in their age from five to twenty days) the cultures were exposed in a water bath to a temperature of 65° C. for two hours, after which they were returned to the incubator for from one to two days, when a tube of fresh bouillon was inoculated from each. The heated cultures were then placed in a refrigerator of about 7° C., where they were kept until used. If the freshly inoculated tubes remained clear the sterility of the heated cultures was considered assured. Before the injection of the sterilized culture fluid the quantity to be used was transferred to a sterile tube and raised to the temperature of the body.

A.—EXPERIMENTS ON GUINEA-PIGS.

The guinea-pigs used weighed from 12 to 16 ounces. They received the culture fluid subcutaneously over the abdomen. The injections were made at short intervals, usually only twenty-four hours intervening between them. To test the immunity acquired the animals were inoculated with 0.1 cc. of a fresh bouillon culture of virulent hog-cholera bacteria in from one to three days after the last injection of the sterilized liquid. In every experiment four guinea-pigs were used; two were treated and two reserved as controls.

The first experiment showed that 15 cc. of a sterilized culture, fifteen days old, injected in fourteen doses, would produce immunity. This very naturally led to the use of cultures of different ages and in smaller quantities. A question of much interest was to determine the number of doses into which it was necessary to divide the required amount to produce a permanent resistance on the part of the animal. To this end several experiments were made, the necessary details of which are given, with the results obtained, in the subjoined table:

I.—*Table showing the immunizing efficiency of sterilized bouillon cultures on guinea-pigs.*

Experiments.	Guinea-pig number—	Quantity of sterilized culture fluid injected.	Number of injections.	Age of cultures when sterilized.	Time from first injection to inoculation.	Date of inoculation with living bacteria.	Control guinea-pigs died in—	Effect on treated guinea-pigs.
		c. c.		Days.	Days.		Days.	
I	1	15	14	15	18	Aug. 26, 1890	8	Recovered.
	2	15	14	15	18	....do	11	Do.
II ...	3	10	10	15	12	Oct. 1, 1890	8	Died in 13 days.
	4	10	10	15	12	....do	9	Recovered.
III ..	5	10	10	10	11	Oct. 24, 1890	6	Do.
	6	10	10	10	11	....do	6	Do.
IV...	7	10	10	10	11	....do	6	Do.
	8	10	10	10	11	....do	6	Do.
V....	9	10	10	20	17	Nov. 11, 1890	7	Killed Nov. 20.
	10	10	10	20	17	....do	7	Died in 18 days.
VI...	11	10	10	5	17	....do	7	Recovered.
	12	10	10	5	17	....do	7	Do.
VII..	13	10	6	10	16	Feb. 4. 1891	6	Died in 13 days.
	14	10	6	10	16	....do	12	Recovered.
VIII.	15	10	4	10	16	....do	6	Died in 12 days.
	16	10	4	10	16	....do	12	Recovered.
IX...	17	10	3	15	12	Nov. 6, 1890	12	Died in 16 days.
	18	10	3	15	12	....do	12	Died in 18 days.
X....	19	8	8	15	11	....do	6	Died in 11 days.
	20	8	8	15	11	....do	6	Recovered.
XI...	21	8	5	10	11	Oct. 24, 1890	6	Died in 7 days.
	22	8	5	10	11	...do	6	Died in 9 days.
XII ·	23	†12	6	....	15	Feb. 4, 1891	6	Died in 6 days.
	24	†12	6	....	15	....do	12	Died in 8 days.

'Control experiment. †Peptonized bouillon.

The immediate toxic effect produced by the subcutaneous injection of the sterilized bouillon cultures was barely observable. The liquid was rapidly absorbed, and in no case did it produce an appreciable irritation. This fact was amply illustrated by the subcutaneous injection of 13 cubic centimeters of a sterilized bouillon culture that had grown for ten days prior to its sterilization. A small guinea-pig was chloroformed and the 13 cc. injected in as many places over the body. This produced a stupefied condition which lasted several hours. The liquid was soon absorbed, leaving the subcutaneous tissue free from indurations. The guinea-pig was inoculated subsequently with 0.1 cc. of a bouillon culture of hog-cholera bacteria. It lived seven days longer than the check and exhibited a large number of tubercle-like bodies beneath the peritoneum.*

It will be observed from the table that in addition to the guinea-pigs that received 15 cc. of the culture all of those that received 10 cc. of five and ten day cultures in ten doses or injections were not seriously affected by the subsequent inoculation of hog-cholera virus. A smaller quantity of the culture fluid or a fewer number of injections were

---

* For a description of this modified form of the inoculation disease, see p. 87.

insufficient to produce immunity, for the inoculation with the strong virus was followed by severe local reaction and the death of a greater or less number of the animals. It is of interest to note that, as anticipated, no resistance was produced by the injection of the peptonized bouillon. (Experiment XII).

The course of the disease in the animals which had acquired a greater or less amount of resistance is exceedingly interesting. The most complete immunity was indicated by the absence of any marked reaction at the place of inoculation. The lesions of those that perished after a certain time exhibited a number of variations from those found in the checks. A slight resistance was characterized by severe local reaction and a greater resistance by certain changes in the organs. Usually a membranous exudate was found over the liver and spleen and the latter was not enlarged. The most marked peculiarity, however, was the formation of a considerable number of small, grayish, tubercle-like bodies. These were usually lenticular in form and varied from 0.25 to 2 mm. in length and from 0.1 to 1 mm. in width. They were distributed in the body as follows: Beneath the peritoneum, especially along the sides of the spinal column; beneath the pleura, more especially over the diaphragm; in the heart muscle, near the apex; and occasionally on the external surface of the abdominal wall. These bodies were firm to the touch. They consisted of a dense infiltration of round cells. Hog-cholera bacteria were present.

In the guinea-pigs which resisted for some days the spleen and liver contained very few bacteria, and frequently no bacilli could be found in cover-glass preparations from them. In similar preparations made from the tubercle-like bodies, crushed with forceps, there invariably appeared a considerable number of hog-cholera bacteria.

The disturbances noticed in the animals after their inoculation with the living bacteria and the more important lesions found upon post-mortem examination are of sufficient interest to deserve a somewhat more detailed statement. In the control animals the lesions found on post-mortem examination were, briefly stated, as follows:

At the point of inoculation a purulent infiltration into the subcutis extending over an area of from 2–4 cm. in diameter; the superficial layer of the subjacent muscle was usually discolored. Inguinal glands enlarged and reddened. Liver contained a greater or less number of necrotic foci. Spleen very much enlarged, dark-colored and friable. Peyer's patches pigmented, often swollen.

Experiment I.—Guinea-pig No. 1 developed a small swelling at the point of inoculation; otherwise both animals appeared perfectly well, and at the time the checks died they had an elevation of temperature of only about one degree. Three days later the temperature was normal. On October 1 they were reinoculated with 0.25 cc. of a bouillon culture of hog-cholera bacteria which produced a very slight local reaction. No elevation of temperature.

Experiment II.—At the time the checks died guinea-pigs Nos. 3 and 4 appeared to be well, if we except a small abscess at the point of inoculation; temperature 0.5° higher than at time of inoculation. Two days later No. 3 appeared to be sick, temperature 101° F. (Normal 102.8° F.) It was found dead on the second morning following. The post-mortem examination showed usual hog-cholera lesions with small grayish tubercle-like bodies beneath the peritoneum on the abdominal wall. Lymphatic glands enlarged. No. 4 remained well.

Experiments III and IV.—Guinea-pigs Nos. 5, 6, 7 and 8 exhibited a slight rise in temperature on the seventh day. A cyst (about the size of a small bean) formed at the point of inoculation, which ruptured and subsequently healed. Nos. 5 and 6 lost no flesh, but Nos. 7 and 8 fell away 0.5 and 1.5 ounces, respectively.

Experiment V.—At the time the checks perished, guinea-pigs Nos. 9 and 10 appeared to be well. Two days later No. 9 was killed for examination. It exhibited a hyperæmic condition of the abdominal viscera. Lymphatic glands enlarged. In No. 10, which died on the 18th day, there was a large local ulcer, exudative peritonitis, and a large number of tubercle-like bodies lying beneath the peritoneum along the left side of the spinal column and beneath the pleura of the left lung. Considerable serum in the pleural and peritoneal cavities.

Experiment VI.—Guinea-pigs Nos. 11 and 12 suffered from severe local reaction; a cyst (containing pus) as large as a walnut formed at the point of inoculation, over which the skin sloughed. They were sick for a few days, but recovered.

Experiment VII.—At the time the check died, guinea-pigs Nos. 13 and 14 appeared to be perfectly well, if we except a slight swelling at the point of inoculation. February 14, both animals were sick. February 17, No. 13 was found dead with a small closed abscess at the point of inoculation. Beneath the peritoneum, on the abdominal walls, and especially along the left side of the spinal column, were scattered a large number of tubercle-like bodies. Lymphatic glands enlarged. Necrosis in liver. A membranous exudate over spleen and liver; spleen not enlarged or discolored. Many of the follicles in Peyer's patches infiltrated. No bacteria discovered in cover-glass preparations from spleen and liver. Covers from the tubercle-like bodies exhibited many hog-cholera bacteria. Guinea-pig No. 14 recovered.

Experiment VIII.—Guinea-pigs Nos. 15 and 16, were very sick at the time the first check died. No. 15 perished with the second check. Usual hog-cholera lesions. No. 16 lived but did not thrive.

Experiment IX.—Guinea-pigs Nos. 17 and 18 suffered from a large abscess at the point of inoculation. Spleens were enlarged and friable, otherwise they presented lesions similar to those found in guinea-pig No. 13. (Experiment VII.)

Experiment X.—Guinea-pig No. 19 exhibited severe local reaction involving subjacent muscle. Spleen not enlarged. Liver, kidneys, and small intestine hyperæmic. The cephalic lobe of right lung in a state of red hepatization. No. 20 apparently perfectly well, excepting a swelling at point of inoculation, which soon subsided.

Experiment XI.—Although the guinea-pigs resisted one and three days respectively, the lesions were practically identical with those found in the checks. Hog-cholera bacteria were found in cover-glass preparations from the spleen and liver.

Experiment XII.—One of the guinea-pigs died with the first check and one died a day before the second check perished. They showed typical hog-cholera lesions.

It is of interest to note the difference in the effect on guinea-pigs between the sterilized cultures of hog-cholera bacteria and solutions or suspensions of the poisonous substances isolated from similar cultures. A perusal of Dr. de Schweinitz's article * shows that the injection of the *sucholotoxin* and *sucholoalbumin*, which he obtained from the cultures, produced severe local reaction, followed in some cases by "ugly ulcers which healed after from ten to fourteen days." The animals appeared ill for several days after the injection, but eventually regained their normal activity. The subsequent inoculation of hog-cholera virus

---

* Medical News, October 4, 1890.

showed that immunity had been induced when the quantity of the toxines injected had been sufficiently large. The guinea-pigs that died
after resisting the disease for a certain number of days after the controls died were examined by one of us. They showed a greater or less
number of tubercle-like bodies beneath the peritoneum. Unfortunately
the number of injections that were made and the quantity of culture
fluid represented in the toxines used are not given, so that further comparisons can not be made.

In the spring and summer of 1893, a few similar experiments were
made on guinea-pigs for the purpose of obtaining immune animals to be
used in other investigations. The readiness with which guinea-pigs
had been made resistant to hog cholera in 1890 led to the supposition
that the same results could be obtained at any time. The following
table indicates, however, that such is not the case. Although an
equally virulent germ was used for making the cultures, and every particular pertaining to the previous experiment adhered to, it was with
much difficulty that the necessary number of guinea-pigs were rendered
immune or at least resistant to the inoculation of virulent hog-cholera
bacteria. The results of these experiments are recorded in Table No. 2,
appended.

II.—*Table showing the immunizing effect of sterilized bouillon cultures of hog-cholera bacteria (1893).*

Experiments.	Guinea-pig number—	Quantity of sterilized culture fluid injected.	Number of injections.	Age of cultures when sterilized.	Time from first injection to inoculation.	Date of inoculation with living bacteria.	Control guinea-pigs died in—	Effect on treated guinea pigs.
		cc.		*Days.*	*Days.*		*Days.*	
XIII	417	10	10	9	29	Jan. 22, 1893	9	Died in 13 days. Emaciated.
	418	12	12	9	29	... do	9	Died in 6 days. Typical hog-cholera lesions.
	419	12	12	9	29	... do	9	Do.
XIV	437	10	10	9	17	Jan. 21, 1893	9	Died in 16 days. Tubercle-like bodies beneath peritoneum.
	438	10	10	9	17	... do	9	Died in 44 days. No lesions, no bacteria.
	439	10	10	9	17	... do	9	Died in 14 days. Tubercle-like bodies beneath peritoneum.
XV	451	12	9	15	20	July 27, 1893	8	Died in 10 days. Typical hog-cholera lesions.
	452	12	9	11	20	... do	8	Died in 11 days. Typical hog-cholera lesions.
	453	12	9	11	20	... do	8	Recovered.
	454	12	9	11	20	... do	8	Died in 18 days. Tubercle-like bodies beneath the peritoneum; hog-cholera bacteria.
XIV *	473	12	13	23	19	Sept. 23, 1893	6	Recovered.
	474	13	13	23	19	... do	6	Do.
	475	13	13	23	19	... do	6	Died in 18 days. Tubercle-like bodies beneath peritoneum and in heart muscle; hog-cholera bacteria.
	476	13	13	23	19	... do	6	Died in 9 days. Typical hog-cholera lesions.
	477	13	13	23	19	... do	6	Died in 10 days† Hog cholera lesions.
	478	13	13	23	19	... do	6	Died in 15 days. Fatty liver, many hog-cholera bacteria.

* In this experiment bouillon containing a small quantity of dextrose was used. The bacteria were
allowed to grow until the fermentation was completed and the liquid became alkaline, when a few more
drops of the dextrose solution was added. This stimulated the growth of the bacteria. When used the
liquid was alkaline in reaction.
† The post-mortem examination of these two animals showed that the injection had been made,
accidentally, into muscular wall of abdomen.

The lesions produced in the guinea-pigs used in the second set of experiments did not differ materially from those found in the animals in the first experiments. Attention is called to the resistance maintained by guinea-pigs Nos. 437, 438, 439, 454, 475, and 478. The cause of death of No. 438 was not determined, as there were no lesions discovered and no bacteria obtained in cultures from the organs. In the other cases the lesions were characteristic of hog cholera, modified in consequence of an increased resistance on the part of the animal. It is probable that Nos. 476 and 477 would have survived if the strong virus-had not been injected into the muscular tissue.

The difference in the outcome between the experiments of 1890 and 1893 can not be satisfactorily explained. It is possible that it was due to the greater susceptibility of the guinea-pigs or to some difference in the cultures. The fact should be stated, however, that the hog-cholera bacteria used were as virulent as those used in 1890. The season may have had some effect in producing the divergent results, as the experiments in 1893 were made during the heat of summer, while the very positive results of 1890 were obtained during a cooler portion of the year.

B.—EXPERIMENTS ON RABBITS.

The certainty with which guinea-pigs could be rendered immune by the use of sterilized cultures in 1890 led to a comparative test on rabbits. Experience with rabbits since 1885 in this laboratory has shown that they are invariably susceptible to hog-cholera bacteria. The experiment on guinea-pigs demonstrated the fact that the principle involved, and for the first time brought out in the pigeon experiments in 1885, was applicable to at least one species of mammals which is very susceptible to hog-cholera bacteria, and thus it answers the charge made by Chantemesse and Widal* that the experiments on pigeons proved nothing because the method failed on susceptible mammals. The experiments on rabbits about to be recorded illustrates very conclusively our inability to draw general conclusions from experiments on one species of animals.

Two small white rabbits received, each, in the subcutaneous tissue, in 11 doses, 20 cc. of sterilized bouillon cultures of hog-cholera bacteria. They were subsequently inoculated, together with two untreated rabbits, with one-eighth cubic centimeter of a bouillon culture of hog-cholera bacteria. They perished with the checks, one in six, the other in seven days after the inoculation. It is instructive to note that *no resistance* to the disease was produced, although each rabbit received one cubic centimeter of sterilized culture fluid for each $1\frac{1}{2}$ ounces of body weight, which quantity was sufficient to produce perfect immunity in guinea-pigs in 1890. The injection of the sterilized culture produced

* Annales de l'Institut Pasteur, II (1888), p. 54.

manifestly no ill effect on the rabbits. The fluid was rapidly absorbed and left no thickening of the subcutaneous tissue.

A third rabbit received 11.4 cc. of sterilized bouillon culture in the ear vein in four injections, at intervals extending over a period of twenty-four days. The first injection (3 cc.), was followed by a temperature elevation of 2° F. On the following day the temperature was above normal. The second injection, of a like quantity, was followed immediately by accelerated respiration and purging (feces blood-stained). The temperature was elevated as after the first injection. The third and fourth injections produced similar but less intense reactions. Unfortunately the rabbit was injured a few days after the fourth injection and hence chloroformed and examined very carefully, but no lesions could be detected.

A fourth rabbit, injected intravenously with 3 cc. of a similar culture, was found dead in three hours.

A fifth rabbit, received in the veins 6.25 cc. in four doses, at intervals extending over a period of thirty-four days. After each injection it exhibited symptoms similar to those already described. It became much emaciated, but otherwise appeared to be well at the time of the test. It died in five days after its inoculation with the virulent culture of hog-cholera bacilli.

EXPERIMENTS WITH STERILIZED AGAR CULTURES OF HOG-CHOLERA BACTERIA.

The agar cultures used in these experiments were prepared as follows: Large tubes (1 inch in diameter) containing 1.5 cc. of agar* were inclined, and the surface inoculated with hog-cholera bacteria. The inoculated tubes were placed in the incubator for several days when the growth on the inclined surface was washed down with sterilized beef broth and transferred by means of a flamed pipette to sterile tubes. A sufficient quantity (usually 10 cc. of sterile beef broth to the growth on four tubes of agar) of liquid was added to make a turbid suspension. When the suspension was to be injected into the veins, it was filtered through sterilized Japanese filter paper, which removed all pieces of agar that might be present. It was sterilized by an exposure to 60° C. for one hour in a water bath, after which it was returned to the incubator for one day. Its sterility was then tested by inoculating fresh tubes of bouillon. If these remained clear the death of the bacteria was considered as positive. The heated cultures were kept, until used, in a cold chamber at a temperature of from 7–9° C.

---

* In the preparation of the agar for this purpose, it was found that the addition of one-eighth to one-tenth per cent of white glue was beneficial in giving to the inclined agar a more firm surface from which the growth could be easily removed. The glue did not interfere with the growth of the hog-cholera bacteria, as shown by parallel cultures on ordinary agar.

A.—EXPERIMENTS ON GUINEA-PIGS.

A few experiments on guinea-pigs with sterilized agar cultures were made. The facts necessary for a full understanding of these test experiments are given in the appended table:

III.—*Table showing the immunizing effect of sterilized agar cultures on guinea-pigs.*

Guinea-pig number—	Quantity of sterilized agar culture injected.	Number of injections.	Time from first injection to inoculation.	Date of inoculation.	Control guinea-pigs died in—	Effect on treated guinea-pigs.
	cc.				Days.	
23	1.5	1	....	....	....	Died in 20 hours after the injection.
24	1.5	1	13	June 30, 1891	6	Died in 8 days.
25	2.0	2	11	....do	6	Died in 9 days.
26	3.5	3	11	....do	6	Died in 15 days.
27	3.5	7	11	June 23, 1893	6	Died in 12 days.
28	3.5	7	11	....do	6	Do.

Guinea-pig No. 23 found to be pregnant, which may explain the fatal toxic effect.

Guinea-pig No. 24 developed a large swelling where fluid was injected, which resulted in the formation of a caseous tumor. The guinea-pig was sick for several days after the injection. It resisted two days. Liver pale, contained necrotic foci. Spleen not enlarged.

Guinea-pig No. 25 exhibited a very large discolored spleen. The injection of the suspension of agar growth produced a small cyst containing a yellowish caseous substance. The local reaction of the hog-cholera virus was very slight.

Guinea-pig No. 26 was apparently perfectly well when No. 25 died, excepting a swelling about the size of a small walnut at the point of inoculation. It lived nine days after the check died. The post mortem showed a membranous exudate over liver and spleen, the latter slightly enlarged; tubercle-like bodies on the external surface of the abdominal wall and heart muscle. No hog-cholera bacteria in cover-glass preparations from the liver.

Guinea-pig No 28 exhibited lesions which did not differ from those found in the check.

Guinea-pig No. 27 contained lesions similar to those found in No. 26.

B.—EXPERIMENTS ON RABBITS.

The rabbits used were young adults. They were kept in large wire cages during their treatment at the Experiment Station of the Bureau. The animals were somewhat carefully watched after the various injections by Dr. F. L. Kilborne, but their temperature was not taken. Usually about six days intervened between the injections. When the rabbits did not thrive a longer period was allowed to elapse between them. The results of the preliminary experiments, together with the necessary details, are given in tabulated form :

IV.—*Table showing the immunizing effect of sterilized agar cultures on rabbits.*

Rabbit number—	Quantity of sterilized suspension injected.	Method of injection.	Number of injections.	Age of cultures when sterilized.	Time from first injection to inoculation.	Date of inoculation with hog-cholera bacteria.	Control rabbits died in—	Effect on treated rabbits.
	cc.			*Days.*	*Days.*		*Days.*	
48	2.75	Ear vein......	3	2	14	June 26, 1891	4	Died in 7 days.
49	.5	...do	1	2				Died a few hours after the first injection.
55	1.5	...do	2	2	14	June 26, 1891	4	Died in 7 days.
62	2.5	...do	3	2				Died after third injection.
63	1.5	...do	2	2				Died after second injection.
64	1.5	...do	2	2				Do.
65	6.25	...do	8	2	80	Sept. 9, 1891	5	Died in 6 days.
58	4.5	Abdominal cavity.	4	2	28	July 20, 1891	6	Do.
59	7.5	...do	6	2	69	Sept. 1, 1891	6	Do.
60	11.5	...do	8	2	78	Sept. 9, 1891	5	Do.
61	17	...do	10	2	90	Oct. 1, 1891	6	Do.
66	6	Beneath the skin.	4	2	28	July 20, 1891	6	Died in 4 days.
67	10	...do	6	2	69	Sept. 1, 1891	5	Died in 6 days.
68	15	...do	8	2	88	Sept. 8, 1891	4	Died in 5 days.
69	22	...do	10	2	98	Oct. 8, 1891	6	Do.

A glance at the table will show that the immunizing power of sterilized agar cultures on rabbits in the quantities used and according to the methods employed was *nil*. The lesions found in the rabbits that perished after the second and third injections are interesting, as they resembled somewhat those produced by the living bacteria. (*See* notes on Nos. 62 and 63 below.) The immediate effect of the injection was marked by depression and apparent illness. The immunity was tested by inoculating the rabbits subcutaneously with 0.05 cc. of a bouillon culture of hog-cholera bacteria (the same as used for the agar cultures) which was sufficiently virulent to destroy rabbits, in the dose given, in from five to seven days. Control rabbits were always used.*

Rabbit No. 49 died in a few hours after the first injection. The examination showed a general hyperæmic condition of the thoracic and abdominal organs. No bacteria.

Rabbits Nos. 48 and 55 showed greater local reaction than that found in the control rabbit.

Rabbit No. 62 died after the third injection of the sterilized agar culture. It had lost no appreciable amount of flesh during the treatment. The examination revealed the following conditions: Several small blood extravasations beneath the serosa in the upper colon and in the rectum. Intestines hyperæmic. In liver a large number of pale reddish areas resembling beginning necrosis. Spleen enlarged, dark colored, friable. Cortex of kidney pale, medullary portion of a bright red color. Lungs hyperæmic, partially collapsed. Heart muscle pale, sprinkled with punctiform hemorrhages. Agar tubes inoculated with pieces of the liver remained sterile.

Rabbits Nos. 63 and 64 perished after the second injection. The lesions were characterized by an enlarged spleen and liver and punctiform hemorrhages in colon and abnormally pale heart muscle. Culture media inoculated from the liver remained sterile.

---

* For the lesions of hog cholera in rabbits, see p. 10.

No. 65 lived one day longer than the check. It exhibited typical hog-cholera lesions.

Rabbits Nos. 58 to 61 and 66 to 69 presented lesions characteristic of hog cholera, and in no case was there any evidence of increased resistance. The lesions were practically identical with those produced in the control rabbits.

These results led to a second series of experiments on rabbits in which the immediate effect of the sterilized cultures was more carefully noted. The suspension of the agar growth was prepared, sterilized and the rabbits injected intravenously as in the preceding experiments:

V.—*Table showing the intravenous injection of sterilized agar cultures of hog-colera bacteria.*

Rabbit number—	Age of culture when sterilized.	Quantity of agar suspension injected.	Number of doses.	Time from first injection to inoculation or death.	Date of inoculation with hog-cholera bacteria.	Results.
	Days.	*cc.*		*Days.*		
107	2	9.5	9	42	....	Very poor, chloroformed for examination.
109	2	7	7	37	Oct. 10, 1891	Died with check.
113	2	1	1	4	....	Poor, chloroformed for examination.
114	2	2	2	7	....	Do.
123	2	6.5	3	23	Oct. 10, 1891	Died 2 days after check.
150	2	7.4	4	39	Nov. 19, 1891	Died with check in 5 days.
110	5	9.8	9	55	....	Died October 30, emaciated and anæmic.
111	5	6.5	6	35	Oct. 10, 1891	Died October 15, with check.

The immediate effect of the sterilized suspension was very marked. Usually the rabbits were purged within thirty minutes after the injection. In a few of the cases the later injections were followed by the evacuation of blood-stained stools.[*] The temperature was elevated from 1 to 3° F. within three hours. The maximum temperature was reached in about twenty hours. It continued above normal for from two to three days. Respiration was accelerated. The rabbits remained very quiet, sat in a crouched position, and usually the eyes were nearly closed. They ate very little the first day after the injection. After the second day they were usually bright and seemed to be quite well. These symptoms were repeated after each injection. The rabbits became emaciated. Those killed for examination revealed no macroscopic lesions, excepting that of anæmia. Those subsequently inoculated with hog-cholera virus, offered no resistance to the disease.

The maximum quantity of the sterilized suspension injected at one time was 3 cc. which was not sufficient, however, to produce death. It

---

[*] It is of interest to note that the severe symptoms produced by the intravenous injection of the agar suspension did not follow the introduction of the filtrate. This was determined by preparing a certain quantity of the suspension, placing it in an incubator for twenty-four hours, filtering it through a bougie of a Pasteur filter and injecting from 1 to 2 cc. of the filtrate into the ear vein of rabbits. The injection was followed by a rise in temperature of from 1.5 to 2.6° F., which subsided within eighteen hours. No other symptoms were detected. The injection of a similar quantity of sterile peptonized bouillon or normal salt solution likewise produced a slight rise of temperature

was observed that when large quantities were injected at first, subsequent injections produced a milder reaction. Tubes of bouillon or agar were always inoculated with bits of tissue from the spleen or liver of the rabbits that died from the effect of the injections. These invariably remained sterile.

C.—EXPERIMENTS ON PIGS.

In order to extend the comparison between the immunizing effect of sterilized bouillon and agar cultures of hog-cholera bacteria, two experiments were made on pigs. The marked resistance obtained in the guinea-pigs with a small quantity of sterilized agar suspension indicated that this substance was more effective than the sterilized bouillon cultures. The effect on rabbits also indicated the presence of a more powerful substance in the agar culture than in the simple bouillon growths, although no immunizing properties were shown.

In the first experiment on pigs a suspension from an agar culture was used that was similar in concentration to that employed in the experiments on other animals. In the second trial the same quantity was injected but the suspension contained the growth from twice the area of agar surface used for the first. The total quantity given each time was divided into several doses, which were injected in different places beneath the skin, on the inside of the thigh, and over the abdomen.

The pigs selected were 3 months old and weighed from 45 to 50 pounds each. They were tested after the treatment by an intravenous inoculation of 6 cc. of a bouillon culture of hog-cholera bacteria. The other facts necessary for an understanding of these experiments are given in a tabulated form.

VI.—*Table showing experiments on pigs with sterilized agar cultures of hog-cholera bacteria.*

Experiment.	Pig number	Quantity of emulsion injected.	Number of injections.	Time from first injection to inoculation with living bacteria.	Date of inoculation.	Controls died in—	Results.
		cc.		*Days.*		*Days.*	
I	59	50	3	34	Aug. 15, 1891		Died in 4 days.
	62	50	3	34	....do		Died in 12 days; extensive ulceration in intestine.
	60	Control			....do	3	
	61	Control			....do	4	
II	67	50	3	26	Oct. 17, 1891		Died in 5 days.
	68	50	3	26	....do		Recovered.
	71	50	3	26	....do		Do.
	69	Control ...			....do	3	
	70	Control ...			....do	3	
	74	Control ...			....do	4	

In the first experiment pig No. 62 lived eight days longer than the checks, and exhibited very extensive ulcerations of the intestinal tract with slight lesions elsewhere in the body. This fact indicated that a certain amount of resistance had been produced which manifested itself by an unusually marked localization of the disease. There were no lesions found at the points of injection of the sterilized agar suspension. In pig No. 67 there were several encysted abscesses in the subcutaneous tissue at the points where the agar suspension was injected. Pigs Nos. 68 and 71 were sick for a few days but fully recovered. The result of the second experiment was sufficiently positive to admit of the inference that two of the three pigs were made immune by the use of the agar cultures. The severity of the test inoculation is indicated by the fact that all of the control pigs and nearly all of the pigs (from the same herd) inoculated simultaneously in connection with certain other experiments carried on at this time perished. It is highly probable that pigs can be made insusceptible to the disease as acquired on the farm by repeated injections of sterilized suspensions of agar cultures. The attending expense of such a process, however, makes it impracticable.

EXPERIMENTS WITH THE BLOOD OF HOG-CHOLERA RABBITS STERILIZED BY HEAT.

In 1890 Selander* published some investigations in which he claimed to have produced, with the sterilized blood of swine-pest-infected pigeons, immunity in rabbits toward the bacillus of swine pest. This bacillus, according to personal observation of one of us, is identical with, or very closely related to, the hog-cholera bacillus, as found in this country.† The bacillus was passed by Selander through a long series of pigeons, and the blood of these pigeons, sterilized by heat, was used to produce immunity in rabbits. His experiments are briefly summarized as follows:

1. *Intravenous injections.*—A rabbit received in an ear vein, in three injections (April 19, May 5, and 9), 3.5 cc. in all of pigeon's blood, from the fifty-first, fifty-ninth, and sixty-sixth passages. The blood was heated for an hour at 57° C. Three days after the last injection the rabbit was inoculated with 0.15 cc. of pigeon's blood from seventy-first passage. It remained well. The control rabbit died in less than thirty-six hours. A second experiment was made, in which 4–5 cc. of sterilized blood was injected in 4 doses. Four days after the last injection the rabbit was inoculated with virulent pigeon's blood of the seventy-third passage. It remained well. The two control rabbits died in less than twelve hours.

2. *Subcutaneous injection.*—A rabbit received 15 cc. of sterilized blood in two doses (March 22, 5 cc.; April 18, 10 cc.). On May 13 it was inoculated subcutaneously with 0.15 cc. of blood of the seventy-first passage. It resisted, while the checks died in less than thirty-six hours.

The discrepancy between the pathogenic action of hog-cholera bacilli and that which Selander describes as characteristic of his culture was sufficient to arouse suspicion. This suspicion is strengthened by the statement of Metchnikoff that his work is a continuation of Selander's, and that he verified the latter's results. It has already been stated

---

* Contribution a l'étude de la maladie infectieuse des porces connue sous les noms de hog-cholera, svinpest, pneumo-entérite infectieuse. Annales de l'Institut Pasteur, IV (1890), p. 545.

† See bulletin on Hog Cholera (1889), p. 181, for a history of the Danish disease.

that we have convicted Metchnikoff of working with the swine plague bacillus and calling it the hog-cholera bacillus in his publications. The evidence is thus very strong that through some inadvertence Selander worked with the swine-plague bacillus. However, as we have not been able to make a study of Selander's bacillus, his work is discussed here under hog cholera, and the reader will note in the following experiments a refutation of Selander's results as regards the hog-cholera bacillus, but a confirmation as regards the swine plague bacillus.

It was found in our experiments that rabbits inoculated with hog-cholera bacilli usually died during the night, and that it was impossible to obtain liquid blood from them post-mortem. For this reason rabbits were inoculated subcutaneously with hog-cholera bacteria, and in the last stage of the disease, presumably only a few hours before death would have occurred, they were killed by bleeding. The blood was collected in sterile tubes, defibrinated, and heated in a water bath for one hour at 58° C. In every case a gelatin roll culture was made prior to the heating with a small loop of the blood. In these innumerable colonies of hog-cholera bacteria developed, which showed that the blood contained enormous numbers of bacteria at the time the animals were killed. The heated blood was either used at once or kept at a low temperature (7–9° C.) until just before it was injected, when the required amount was heated to the body temperature. Previous experiments[*] had demonstrated the fact that hog-cholera bacteria are destroyed by exposing them to 58° C. for fifteen minutes. The certainty with which the bacteria are destroyed by such an exposure to heat led to the use of the blood in some cases immediately after it was heated, in order that no change might be produced by its standing. Its sterility, however, was tested in every case by subcultures in bouillon. These cultures invariably remained clear.

The rabbits from which the blood was obtained were inoculated subcutaneously with 0.1 cc. of a fresh bouillon culture of hog-cholera bacteria. A brief history of each of these animals is appended:

Rabbit No. 336, inoculated with virulent hog-cholera bacteria August 11, 1893; killed by bleeding August 15 (four and one-fourth days). The blood, after being sterilized, was kept at a temperature of from 7–8° C. for 20 hours, when it was used for the first injection.

Rabbit No. 310, inoculated August 15; bled August 19 (four and one-fourth days). The blood was used for the second injection immediately after it had been sterilized.

Rabbit No. 344, inoculated August 19; bled August 25 (six and one-fourth days). A portion of the blood was used immediately after it had been sterilized for the third injection. The remainder was kept at a temperature of from 7–8° C. for two days and then used for the fourth injection.

Rabbit No. 350, inoculated August 26; bled August 30 (four and one-fourth days). The blood was used immediately after its exposure to the heat for the fifth injection.

The post-mortem examination of the rabbits furnishing the blood showed, without exception, an enlarged spleen and necrosis in the liver. The three first were unable to stand when they were killed.

---

[*] Second Annual Report of the Bureau of Animal Industry (1885), p. 215.

The fourth rabbit (No. 350) was very sick, but evidently would have lived a few hours longer than the others. The number of colonies which developed in the gelatin roll cultures, made from the blood immediately after its defibrination, indicated that there were as many bacteria present in the blood of these animals as there are in the blood of rabbits which are allowed to die.

In this experiment with the sterilized blood young adult rabbits were used. The number of rabbits employed, quantity of blood injected, and the result of the test inoculation with hog-cholera bacteria are given in the following table:

VII.—*Tables showing the immunizing effect of sterilized blood of rabbits suffering from hog cholera.*

1. STERILIZED BLOOD INJECTED INTRAVENOUSLY.

Source of blood.	Rabbit No. 336.	Rabbit No. 310.	Rabbit No. 344.	Rabbit No. 344.	Rabbit No. 350.	Total amount injected.	Inoculated subcutaneously with 0.1 cc. of a bouillon culture hog-cholera bacteria.	Results.
Date of injection..........	Aug.16.	Aug.19.	Aug.25.	Aug.27.	Aug.30.			
Rabbit—	cc.	cc.	cc.	cc.	cc.	cc.		
No. 337.....	0.75	1.30	1.30	1	2.05	6.40	Sept. 3. 1892...	Died Sept. 8, 1893.
No. 338.....	1	1.30	1.30	1	1.80	6.40	....do	Do.
No. 339.....	1.40	1.30	1.30	1	1.40	6.40	....do	Died Sept. 9, 1893.
No. 348.....	....	....	1	1.25	1.40	3.65	....do	Do.
No. 349.....	....	....	1	1.25	1.40	3.65	....do	Died Sept. 7, 1893.
No. 355.....	Check.	....	....	....	....	....	....do	Died Sept. 8, 1893.

2. STERILIZED BLOOD INJECTED SUBCUTANEOUSLY.

Source of blood.	Rabbit No. 336.	Rabbit No. 310.	Rabbit No. 344.	Rabbit No. 344.	Rabbit No. 350.	Total amount injected.	Inoculated subcutaneously with 0.1 cc. of a bouillon culture hog-cholera bacteria.	Results.
Date of injection..........	Aug.17.	Aug.19.	Aug.25.	Aug.27.	Aug.30.			
Rabbit—	cc.	cc.	cc.	cc.	cc.	cc.		
No. 340.....	2.8	2.6	2.6	2	5	15	Sept. 3,1893...	Died Sept. 7, 1891.
No. 343.....	2.8	2.6	2.6	2	5	15	....do	Do.
No. 355.....	Check.	....	....	....	....	....	....do	Died Sept. 9, 1893.

At the same time a number of guinea-pigs were treated subcutaneously with sterilized blood from the same rabbits. This experiment is summarized in the subjoined table:

VIII.—*Table showing the immunizing effect on guinea-pigs of sterilized blood of rabbits suffering from hog cholera.*

Source of blood.	Rabbit No. 336.	Rabbit No. 310.	Rabbit No. 344.	Rabbit No. 344.	Rabbit No. 350.	Total amount injected.	Inoculated sub cutaneously with 0.2 cc. bouillon culture of hog-cholera bacteria.	Results.
Date of injection..........	Aug.16.	Aug.19.	Aug.25.	Aug.27.	Aug.30.			
Guinea-pig—	cc.	cc.	cc.	cc.	cc.	cc.		
No. 279	0.5	1.3	1.3	....	....	3.1	Sept. 3, 1893...	Died Sept. 9.
No. 280	1	1.3	1.3	1	....	4.6	....do	Do
No. 272	....	....	....	1.25	1.4	2.65	....do	Died Sept. 10.
No. 283	....	....	....	1.25	1.4	2.65	....do	Do.
No. 288	Check.	....	....	....	....	....	....do	Do.
No. 289	...do ..	....	....	....	....	....	...do	Do.

It will be observed that even a slight degree of immunity was not produced. While the results differ from those obtained by Selander they are in conformity with those heretofore obtained in this laboratory with hog-cholera bacteria.

The toxic effect produced in the rabbits by the intravenous injection of the sterlized blood was very slight. The temperature rose from 1° to 1.8° F. after the first injection, but subsequently no elevation was detected. This was true of the subcutaneous injections in both rabbits and guinea-pigs. At no time did the animals refuse food, although they appeared to be unusually quiet for a short time after the first injection.

EXPERIMENTS WITH THE BLOOD SERUM OF RABBITS MADE INSUSCEPTIBLE TO FATAL DOSES OF HOG-CHOLERA BACILLI INTRODUCED UNDER THE SKIN.

These experiments were undertaken primarily to examine into the claims of Metchnikoff concerning the efficacy of blood serum from immune rabbits as a preventive and curative of hog cholera artificially induced in rabbits by inoculation.

In the course of these investigations a culture of the "hog-cholera" bacillus used by Metchnikoff in his work was asked for and kindly sent for examination. The study of this culture showed that Metchnikoff had not had the hog-cholera bacillus at all, but the swine-plague bacillus.* How this unfortunate mistake had come about it is not worth while to speculate upon here. It is sufficient to state that the persistence with which European observers have at first denied the existence of one or the other of these pathogenic forms and then confounded them, one with the other, is almost entirely responsible for the confusion which, from their point of view, exists in this subject.

The results obtained by Metchnikoff at first inexplicable are thus easily interpreted, and they agree entirely with the results detailed in

---

* Metchnikoff kindly sent the blood of a guinea-pig inoculated with the bacteria in a sealed glass tube. Two of these tubes were examined and in both the same bacteria were found in a state of purity. Cultivated on various substrata they presented all the characters of swine-plague bacteria as found in this country. The bouillon growth is always feeble, the agar growth quite vigorous, forming a grayish membrane on the agar surface. In bouillon and on agar the bacteria appear as very minute cocci, having at times a marked dancing (Brownian?) motion utterly unlike the rapid spontaneous movement of hog-cholera bacilli. The characteristic polar stain so commonly seen in preparations from the tissues of rabbits is only rarely recognizable in cultures. On agar plates prepared in closed Petri dishes the colonies emit a peculiar, disagreeable odor very characteristic of the whole group of swine-plague bacteria. In tubes of gelatin or on plates development remains nearly always invisible to the eye. Milk remains unchanged. On potato there is no visible growth. In glucose bouillon gas is not set free.

The pathogenic action as shown on rabbits was equally peculiar to the swine-plague group.

November 22, 1892. Male rabbit, 2⅛ pounds, was inoculated subcutaneously with two drops of blood from one of the original sealed pipettes.

November 26. Found dead. At site of inoculation there was a small area of puru-

this article, provided we substitute the words "swine-plague bacillus" wherever he has used the words "hog-cholera bacillus." A description of his experiments is therefore omitted here and given on page 74, together with the work on the protective action of the blood serum of rabbits made insusceptible to swine plague.

In our experiments blood serum from rabbits B, No. 205 and No. 232, was used. (*See* p. 43.)

The rabbits were secured and anæsthetized with chloroform or ether. A carotid or a femoral artery was laid bare with proper aseptic precautions and a sterile glass canula inserted. The blood was allowed to flow into sterile pear-shaped glass bulbs and kept over night in a temperature of about 70° F. The clear serum was withdrawn with sterile pipettes on the following day and kept in a refrigerator.

The following table summarizes the first experiment:

IX.—*Table showing immunizing efficiency of serum from rabbit B.*

Blood drawn ...	Dec. 2, 1892.*	Dec. 7.	Dec. 14.		Total quantity of serum injected.	Inoculation with 0.15 cc. virulent bouillon culture.	Result.
Date of injection of serum.	Dec. 3.	Dec. 5.	Dec. 8.	Dec. 9.	Dec. 15.	Dec. 16.			
Rabbit—	cc.	cc.	cc.	cc.	cc.	cc.	cc.		
No. 403.....	1.4	1.4	1.4	......	1.4	1.4	7	Dec. 29......	Dead Jan. 3.
No. 404.....	1.4	1.4	1.4	......	1.4	1.4	7	Dec. 29......	Do.
No. 378.....	......	......	1.4	1.4	1.4	1.4	5.6	Dec. 29......	Do.
No.405 (control)	......	......	......	......	......	......	......	Dec. 29......	Do.
Guinea pig, 354.	1.4	1.4	1.4	......	1.4	1.4	7	Dec. 29......	Dead Jan. 13.

* Last inoculation with virulent culture September 15, 1892.

From this table it will be noted that there was no apppreciable amount of resistance acquired by the rabbits during the treatment with blood serum. The guinea-pig lived about eight days longer than the usual period. Unfortunately no control was used with this animal as the use of the guinea-pig was rather an after-thought in this experiment. It is not necessary to give the autopsy notes of these animals, as they present nothing unusual.

lent infiltration with surrounding hemorrhages and ecchymoses. Extravasations on omentum and on lower portion of large intestine. Spleen large, dark, and firm. In left lung 5 small hemorrhagic foci. In stained cover-glass preparations of spleen pulp a large number of bacteria showing the polar stain very well. In heart's blood very few. Cultures from blood and spleen positive.

January 10, 1893. A rabbit received into an ear vein 0.12 cc. of a bouillon culture twenty-four hours old. This culture was derived from the blood in the sealed pipette through agar plates and agar subcultures from a colony. The rabbit was found dead next morning with hyperæmic lungs and punctiform ecchymoses under serosa of large intestine. Immense numbers of polar-stained bacteria in the spleen. This organism has been kept growing up to the present with no change in its characters beyond a gradual decline of virulence.

The following table summarizes a later experiment with the serum of rabbits Nos. 205 and 232:

X.—*Table showing immunizing efficiency of serum from rabbits 205 and 232.*

Blood from	No. 205.*	No. 232.†	No. 205.	Total quantity of serum injected.	Inoculation with virulent bouillon culture Feb. 13.	Results.
When drawn	Jan. 4. 1893.	Jan. 11.	Jan. 23.			
Date of injection of serum	Jan. 5.	Jan. 6.	Jan. 12. Jan. 13. Jan. 25. Jan. 28.			
	cc.	cc.	cc. cc. cc. cc.	cc.	cc.	
Rabbit—						
No. 416	1.3	1	1 1.4 1.4 0.5	6.6	0.06	Dead Feb. 18.
No. 413	1.3	1	1 1.4 1.4 0.5	6.6	0.06	Do.
No. 415 (control)					0.06	Dead Feb. 20.
Guinea pig—						
No. 365	1.3	0.75	1 1.4 1.4	5.85	0.09	Dead Feb. 22.
No. 366	1.3		1 1.4 1.4	5.1	0.09	Dead Feb. 20.
No. 367 (control)					0.09	Do.
No. 369 (control)					0.09	Dies Feb. 20 (noon).

* Last inoculation with virulent culture of hog-cholera bacilli was made September 15, 1892.
† Last inoculation July 29, 1892.
The guinea-pigs weighed about 1.5 pounds, the rabbits 3 to 3.5 pounds.

In connection with the blood-serum injections a curious effect on the guinea-pigs was noticed. In No. 365 a bluish discoloration 1 to $1\frac{1}{2}$ inches in diameter appeared at the place of injection (abdomen), with more or less firm infiltration in the subcutaneous tissue. At the time of the test inoculation the lesion had disappeared.

In No. 366 a similar infiltration, without discoloration, appeared after two injections, one of which became ulcerated. This also had healed February 13.

When first observed it was suspected that perhaps a few hog-cholera bacilli had been in the animals from which the serum was drawn. This supposition was disposed of by the fact that no infiltration or reaction of any kind followed the injection in the more susceptible rabbits. To what other causes this phenomenon may be referred to, inquiries have not yet been made. It may have been due either to the irritating action of the rabbit's serum *per se* or to some change in the serum brought about by the acquired immunity.

This experiment may be objected to, as illustrating the absence of any protective substance in the blood in appreciable quantity, on two grounds—the long interval between the drawing of the blood and the last virulent inoculation, and the long interval between the last blood-serum injection and the test inoculation (sixteen days). These objections are to a certain extent well founded. There were, however, sufficient reasons for the delay. The hog-cholera bacillus is not eliminated from the organs of immune rabbits for *at least* a month after inoculation and it probably persists much longer. Hence any haste in the collection of the blood may either miss the period of maximum protective power or it may lead to the collection of blood containing living hog-cholera bacilli. The test inoculation following the blood-serum treatment was delayed to give the indurations and ulcers produced by the blood serum in guinea-pigs sufficient time to heal.

The absence of any retarding effect of the blood serum on the disease in rabbits under the conditions of the experiment shows that any appreciable increase of resisting power is not likely to be obtained by increasing the quantity of the serum. This result is in entire accord with the negative outcome of the other immunizing methods applied to rabbits. With guinea-pigs the result is somewhat different. There seems to have been a slight degree of immunity induced in the guinea-pig used in the first blood-serum experiment. This also is in accord with the sterile bouillon experiment and the blood-serum experiment on guinea-pigs, described below.

DOES THE BLOOD OF IMMUNE RABBITS POSSESS ANY BACTERICIDE OR ANTITOXIC POWER?

Only one trial was made. Some preliminary experiments carried out by Dr. C. F. Dawson in this laboratory had shown the absence of any bactericide action of the blood serum of immune rabbits on hog-cholera bacilli. The blood serum used by him was obtained from one or the other of the rabbits Nos. 205, 232, and B.

In January of 1894, rabbit B was killed and a small quantity of blood-serum obtained, which was used as follows :

A definite quantity of a bouillon culture of hog-cholera bacilli, twenty-four hours old, was added by means of a carefully graduated drop pipette to a definite quantity of serum and the mixture injected into rabbits subcutaneously after a certain length of time, as indicated in the subjoined table:

XI.—*Table showing results of injection of rabbits with mixture of hog-cholera bacilli and serum.*

Rabbit.	Serum mixture contains—		Time elapsing between mixing serum and bacilli and inoculation into rabbit.	Date of inoculation.	Result.
	Of bouillon culture of hog-cholera bacilli.	Serum.			
	cc.	cc.	Hrs. min.		
No. 127	.033	0.5	0 15	Jan. 10, 1894	Dead January 17.
No. 128	.033	0.5	1 48	....do	Dies January 15, noon.
No. 126	.033	0.5	24 00	Jan. 11. 1894	Dies January 16.
No. 84	.033	*0.5	....	Jan. 10, 1894	Dies January 13.

* Sterile bouillon.

The cause of death in these animals was determined by a careful autopsy, by cultures and by the examination of cover-glass preparations of the spleen pulp. There seems to have been a slight retardation of death in favor of the fifteen-minute serum mixture. However, the period of the disease produced by it was within the normal limit, while

that of the control rabbit was shorter than is usually the case. Any
protective action of the serum mixture can not be deduced. It should
be stated that the rabbit furnishing the serum had not been inoculated
with hog-cholera bacilli since September 15, 1892.

EXPERIMENT ON GUINEA-PIGS WITH BLOOD SERUM FROM GUINEA-PIGS IMMUNIZED AGAINST HOG-CHOLERA BACTERIA.

A single experiment was made on guinea-pigs to test the immunizing efficacy of the blood serum from guinea-pigs previously made
immune to hog cholera by artificial means. The method of immunizing the animals from which to obtain the serum was the subcutaneous
injection of sterilized bouillon culture of hog-cholera bacteria. A
detailed account of the immunization of these guinea-pigs is given on
p. 50, under the numbers 474, 473, and 453. No. 453 had been inoculated
twice with hog-cholera bacteria, the other two once only.

Guinea-pig No. 474 was bled twenty-four days after the test inoculation and
seventeen days after the control guinea-pig had died.

Guinea-pig No. 473 was bled thirty-seven days after the test inoculation and
thirty-one days after the control had died.

Guinea-pig No. 453 was bled thirty-eight days after the second test inoculation
and thirty-two days after the death of the control animal.

At the time the guinea-pigs were bled they were in excellent condition. After they had been bled and killed a careful examination
showed that they were free from any lesions of disease. As a pre-
cautionary measure, however, tubes of bouillon were inoculated with
several large loops of the blood and with a large piece of the spleen of
each animal. Stained cover-glass preparations from the various
organs showed no bacteria. The tubes of bouillon inoculated with the
blood remained clear, but those inoculated with the spleen developed
into pure cultures of hog-cholera bacteria on the second day. One
rabbit was inoculated with each of these cultures. All died of hog
cholera in the usual time.

The length of time during which the hog-cholera bacteria remained
alive in the spleen of these animals and the unchanged virulence are
· facts worthy of notice. The permanent sterility of the bouillon tubes
inoculated with the blood would indicate that the bacteria were not
present in this fluid at the time the guinea-pigs were killed, and that
the serum used was free from living bacteria.

To obtain the blood from the immune guinea-pigs the animals were
etherized, the hair removed from the neck, and the skin thoroughly
disinfected. The carotid artery was then exposed with sterile instru-
ments and the animal held in such a position that the blood from the
incised artery could run into a sterile funnel and thence into a sterilized pear-shaped glass bulb. About 25 cc. of blood was obtained from
each animal. This was placed in a temperature of about 8° C. From
4 to 7 cc. of clear serum was obtained from each animal. The injection

of guinea-pigs with the serum, the subsequent test inoculation, and its result are summarized in the subjoined table:

XII.—*Table showing immunizing effect of blood serum from immune guinea-pigs injected subcutaneously.*

Source of serum.	Guinea-pig—No. 474.	Guinea-pig—No. 474.	Guinea-pig—No. 473.	Guinea-pig—No. 473.	Guinea-pig—No. 453.	Total quantity of serum injected.	Inoculated with 0.1 cc. bouillon culture of hog-cholera bacteria.	Results.
Date of injection.	Oct. 20, 1893.	Oct. 25, 1893.	Nov. 2, 1893.	Nov. 6, 1893.	Nov. 9, 1893.			
Guinea-pig—	cc.	cc	cc.	cc.	cc.	cc.		
No. 1	1	1	1	1	2	6	Nov. 13, 1893 ..	Died Nov. 27, 1893.
No. 4	1	1	1	1	2	6	....do	Died Dec. 7, 1893.
No. 2				1.5	3	4.5	....do	Died Nov. 24, 1893.
No. 5					2	2	....do	Do.
No. 8						check	....do	Do.

The results obtained in this experiment are more positive than those obtained by the use of sterilized rabbit's blood. While the injection of 2 cc. of the serum produced no effect the, use of 6 cc. produced an appreciable resistance. The large number of tubercle-like bodies found in guinea-pig No. 2 may be taken as an indication of partial immunity, although it died with the check. No. 4 survived the inoculation for twenty-four days. The amount of resistance induced by the blood serum was no greater in proportion to the quantity used than that obtained with the agar cultures sterilized by heat.

SWINE PLAGUE

The swine-plague bacteria used in this series of experiments are described in the Special Report on the Cause and Prevention of Swine Plague, Department of Agriculture, 1891 (p. 57). They were sufficiently virulent to destroy rabbits inoculated subcutaneously with 0.001 cc. of a fresh bouillon culture in less than twenty hours. The following experiments are similar in detail to those made with hog-cholera bacteria.

EXPERIMENTS WITH STERILIZED BOUILLON CULTURES OF SWINE-PLAGUE BACTERIA.

In these experiments young adult rabbits were used. The sterilized bouillon cultures were prepared and test inoculations were made in accordance with the methods given in the hog-cholera experiments. Control rabbits were always used. These died without exception within twenty-four hours. On this account they are omitted from the table giving the details of the treated rabbits.

The immediate, toxic effect of the injection of the sterilized bouillon cultures varied according to the method of injection. The rabbits that received the cultures in the peritoneal cavity sat very quietly for a few hours, but otherwise they appeared to be well. The intravenous injection produced a more marked depression. The animals sat in a crouched position, refused food, and offered no resistance to handling. The

22275—No. 6——5

temperature rose 1 to 1.5° F. Usually the effect was noticeable for about twenty-four hours, after which time the animals appeared to be in their normal condition. These well-marked disturbances occurred after the first injection only. Subsequent injections produced very little and often no appreciable effect. A summary of the facts in this experiment is given in the following table:

XIII.—*Table showing results of experiments with sterilized bouillion cultures of swine-plague bacteria.*

Rabbit—	Date of first injection.	Method of injection.	Total quantity of sterilized culture injected.	Number of injections.	Test inoculation with living swine-plague bacteria.	Results.
			cc.			
No. 12....	Dec. 26, 1890	Ear vein...	7	5	Jan. 10, 1891	Died Jan. 19. Severe local reaction and peritonitis.
No. 13....	Jan. 26, 1891	....do	7	5	Feb. 24, 1891	Died Feb. 26.
No. 14....	Apr. 24, 1891	....do	8	3	May 19, 1891	Died May 20.
No. 15....	....do ,	....do	12	4	May 26, 1891	Died June 3. Severe local reaction, pleuritis, and peritonitis.
No. 16..:.	....do	....do	16	5	June 5, 1891	Died June 11, 1892. Very slight local reaction, which healed; large closed abscess in abdominal cavity.
No. 17....	....do	....do	16	5	....do	Recovered. Killed for examination Mar. 24, 1892.
No. 18....	Mar. 10, 1891	Abdominal cavity.	20	4	Mar. 31, 1891	Died Apr. 1.
No. 19....	....do	....do	20	4	... do	Do.
No. 20....	....do	....do	40	8	Apr. 17, 1891	Died Apr. 18.
No. 21....	....do	....do	40	8	....do	Recovered. Severe local reaction. Killed for examination Feb. 9, 1892.

The foregoing table shows that the injection of the sterilized cultures into the veins produced a greater degree of resistance than the injection into the abdominal cavity. It will be found from the appended post-mortem notes that the degree of resistance on the part of the treated animal was accompanied by variations in the course of the disease, so that when a rabbit resisted the inoculation of virulent swine-plague bacteria for a number of days it exhibited lesions similar to those found in rabbits that died from an inoculation of an attenuated swine-plague germ which required several days to destroy life. These variations are especially well marked in rabbits Nos. 12 and 15. The effect of the inoculation in rabbit No. 16 is of special interest, as the animal showed no disturbance for several months after the inoculation, but died more than a year later. This variation in the course of the disease will be more fully illustrated in the experiments with agar cultures. The interesting points in the post-mortem notes of these rabbits are appended. Attention is particularly called to those of rabbits Nos. 12, 15, and 16.

Rabbit No. 12 lived three days after inoculation. There was quite a severe purulent infiltration at the place of inoculation. Surrounding this the blood vessels were much injected; about 4 cm. from the point of inoculation toward the knee-fold a small mass of purulent substance. On the cæcum a small quantity of a grayish, pasty exudate was collected into bands and shreds which contain leucocytes and an innumerable number of swine-plague bacteria. Under the serosa of the small intestine, excepting duodenum, are numerous blood extravasations of irregular outline. Lungs œdematous and partially collapsed. A large number of polar-stained bacteria in preparations from the spleen and blood. A cover-glass touched to the pleura on the diaphragm and stained showed a few swine-plague bacteria.

Rabbits Nos. 13 and 14 died within forty-eight and twenty-four hours of acute swine plague.

Rabbit No. 15 lived eight days. At the point of inoculation the subcutis was infiltrated with a purulent exudate extending over an area 6 cm. in diameter; subjacent muscles necrosed; surrounding blood vessels injected. Over intestines were a few fine shreds of exudate. Liver hyperæmic. Spleen very dark. Kidneys pale. Right lung hyperæmic; pleura, both parietal and visceral, covered with a thin cellular exudate. In cephalic lobe two foci about 1 cc. in diameter in a state of grayish consolidation. The left pleural cavity lined with a quite thick membranous exudate which covers the entire surface of the lung. On the dorsal surface of the principal lobe an area 2 cm. in diameter and extending through the lung was consolidated, of a yellowish-gray color, apparently necrosed; the remaining portion of the principal lobe hyperæmic; cephalic lobe collapsed. Pericardium covered with a thin cellular exudate. Heart muscle pale. The exudate contained an innumerable number of swine-plague bacteria. Very few bacteria in cover-glass preparations from the spleen.

Rabbit No. 16 lived one year and six days after inoculation. This rabbit appeared to be perfectly well for several months after the inoculation. In the spring of 1892 it was observed to be emaciated to a considerable degree. It died June 11, 1892. The post-mortem examination showed the rabbit to be very poor. The abdominal cavity contained a very large tumor extending from the pubes cephalad and covering the kidneys and crowding the intestines up under the ribs. The cæcum was lying over the stomach. The tumor was found to be beneath the peritoneum, the blood vessels of which were injected. The ovaries, Fallopian tubes, ureters, and rectum were plainly visible on the convex surface of the tumor. An incision showed the tumor to consist of an outer rather firm wall about 2 mm. thick, containing a whitish substance resembling fine wheat-flour paste in consistency. A microscopical examination showed it to consist of broken-down pus cells and fine granules. The tumor when removed weighed 900 grams or nearly one-half the entire weight of the animal (total weight 1,980 grams). The liver, spleen, and kidneys were small and firm. Heart and lungs apparently normal. A tube of bouillon was inoculated and a series of agar plate cultures made from the contents of the tumor. These developed into pure cultures of swine-plague bacteria. The virulence of the germ was tested on rabbits with the following results:

June 15. *Rabbit No. 286* was inoculated subcutaneously with 0.30 cc. of a bouillon culture. June 18, rabbit found dead with lesions characteristic of attenuated swine plague, i. e., severe local infiltration and peritonitis.

June 30. *Rabbit No. 316* was inoculated subcutaneously with 0.25 cc. of a bouillon culture made from the blood of rabbit No. 286. July 1, rabbit found dead. Innumerable swine-plague bacteria in the organs.

July 2. *Rabbit No. 315* was inoculated subcutaneously with 0.01 cc. of a bouillon culture from rabbit No. 316. It died within twenty-four hours.

These inoculations were sufficient to restore the virulence of the bacteria inoculated into rabbit No. 16, and to demonstrate the identity of the bacteria obtained from the closed abscess with the bacteria inoculated a year previously.

Rabbit No. 17. This animal was killed for examination March 24, 1892. It was found in good condition. No bacteria were found either in cover-glass preparations or in cultures from the organs.

Rabbits Nos. 18, 19, and 20 died from the acute, septicæmic form of swine plague. *Rabbit No. 21,* killed February 9, 1892, nearly ten months after its inoculation. Very much emaciated. Beneath the skin over abdomen several small cysts containing pus. Projecting into abdominal cavity from the dorsal wall is a tumor about 8 cm. in diameter. Upon inspection it was found to consist of a grayish, pasty, purulent substance inclosed in a sac having a wall about 2 mm. in thickness. A tube of agar inoculated with a loop of the heart's blood remained clear. A tube of bouillon inoculated from the contents of the tumor developed into a pure culture of swine-plague bacteria. A rabbit inoculated subcutaneously with 0.25 cc. of this culture died within twenty-four hours.

EXPERIMENTS WITH STERILIZED SUSPENSIONS OF AGAR CULTURES OF SWINE-PLAGUE BACTERIA.

In the preparation of the suspensions of the agar culture the same method was pursued as in the hog-cholera experiments. The cultures varied in age from two to five days. The surface growth was diluted with bouillon and filtered through sterilized Japanese filter paper, to remove all masses of agar and clumps of bacteria. The filtrate had a grayish, turbid appearance. The suspension was distributed in large sterile test tubes, closed with cotton plugs and sterilized by heating in a water bath for one hour at 60° C. The sterility of the heated cultures was determined by means of subinoculations in bouillon. Control rabbits were used in all experiments, but are omitted from the table. As in the preceding experiment, they invariably succumbed to the subcutaneous inoculation within twenty-four hours. The rabbits used in the experiment were young adults. The details of this experiment are given in the following table:

XIV.—*Table showing results of experiments with sterilized suspensions of agar cultures of swine-plague bacteria.*

Rabbit—	Date of first injection.	Method of injection.	Total quantity of sterilized suspension used.	Number of injections into which it was divided.	Inoculated with living swine-plague bacteria.	Results.
			cc.			
No. 38....	May 4, 1891	Abdominal cavity.	4. 5	3	May 19, 1891	Died May 25, 6 days after inoculation.
No. 35....	....do	....do	7. 5	4	May 26, 1891	Died Feb. 2, 1892, 9 months after inoculation.
No. 37....	....do	...do	7. 5	4	....do	Recovered.
No. 36....	....do	....do	7. 5	4	....do	Do.
No. 51....	June 9, 1891	....do	3	2	June 19, 1891	Died June 22, 3 days after inoculation.
No. 50....	....do	...de	3. 5	3	June 29, 1891	Recovered.
No. 52....	....do	...do	12	4	....do	Do.
No. 53....	....do	...do	12	4	....do	Do.
No. 23....	....do	Ear vein...	1. 5	2	June 19, 1891	Do.
No. 54....	....do	...do	2	3	June 29, 1891	Do.
No. 76....	June 29, 1891	....do	2. 5	3	July 7, 1891	Died July 8, with control rabbit.
No. 77....	....do	...do	2. 5	3	....do	Do.
No. 130...	Oct. 1, 1891	....do	2	1		Died Oct. 2, from the effect of treatment.
No. 129...	....do	...do	3. 5	3	Oct. 15, 1891	Died Nov. 4, 20 days after its inoculation.
No. 152...	Oct. 16, 1891	....do	5. 5	3	Nov. 19, 1891	Recovered.
No. 79....	June 29, 1891	Subcutis...	4	3	July 7, 1891	Died July 8, with control rabbit.
No. 78....	....do	...do	8. 5	5	July 18, 1891	Recovered.

The injections of sterilized suspension of swine-plague bacteria were repeated at intervals of from two to six days according to the effect produced. Time was allowed for the normal condition of the animal to become restored after each injection before the latter was repeated. In case of the intravenous injection this period was much longer than in the other cases. The test inoculation was made in from two to six days after the last injection.

The immediate or toxic effect of the agar suspension was more severe than that produced by the bouillon culture. The abdominal injection was followed by a slight indifference on the part of the rabbits when handled and their refusal of food. The intravenous injection, however, produced a marked effect. There was an elevation of the temperature of from 0.5° to 2° F. within two hours. Respiration accelerated. Eyes usually more or less closed. In from twelve to eighteen hours the temperature reached 104.5° to 105.6° F. which elevation continued for about twenty-four hours, when it rapidly subsided to the normal. During this time the rabbits refused food, offered no resistance to handling, and the fur had a ruffled appearance. In some cases the injection was followed within twenty minutes by a copious evacuation of the bowels. Subsequently the rabbits appeared to suffer from tenesmus for a shorter or longer time. The second injection produced the same symptoms. After the third injection the reaction was very slight and often inappreciable.

The subcutaneous injections produced no appreciable constitutional symptoms. There was a slight infiltration of the subcutis at the points of injection which disappeared in a few days.

It is of interest to note that in rabbits Nos. 38 and 51 which received small quantities of the suspension there was very severe local reaction after the inoculation with the strong virus followed by peritonitis and death. In No. 50 there was severe local reaction with recovery. In those that received 7.5 cc. the local reaction was less severe and recovery followed in two of the three cases. No. 35 is extremely interesting from the fact that the rabbit lived nearly ten months. In the two rabbits which received 12 cc. each, there was scarcely any local reaction and the animals remained apparently perfectly well.

It is difficult to understand the effect of the intravenous injections. Rabbits Nos. 23 and 54 resisted the inoculation of the strong virus, though they had received only a small quantity of the suspension, while Nos. 78 and 79 offered no resistance, although they had received a greater quantity. The subcutaneous injection of the agar suspension was equal to the abdominal injection in efficiency. The rabbits that perished after showing a marked resistance were affected with lesions that are of sufficient interest to be briefly recorded:

Rabbit No. 38 resisted sixty days. A purulent infiltration at point of inoculation over an area 3 cm. in diameter; subjacent muscle pale. Spleen slightly enlarged. Liver congested. The pleura of both lungs and parietes covered with a grayish friable exudate consisting of cells and bacteria. Exudative pericarditis. Lungs deeply hyperæmic, only partially collapsed.

Rabbit No. 51 resisted three days. A purulent infiltration in the subcutis at point of inoculation over an area of about 2 cm. in diameter. Over abdominal viscera a grayish exudate composed of round cells and swine-plague bacteria.

Rabbit No. 35 resisted for nearly nine months. Rabbit very much emaciated. At point of inoculation a closed abscess about the size of a horse-chestnut, containing pus of a pasty consistency. Thoracic and abdominal organs pale; brain and spinal cord apparently normal. Culture media inoculated from the local abscess and blood remained clear. The time which had elapsed (nearly ten months) after the inoculation was sufficiently long to admit of other causes, but it seems more likely on account of the local abscess that the swine-plague inoculation was primarily responsible.

Rabbit No. 129 resisted twenty days. It exhibited a small abscess at the point of inoculation. The abdominal cavity contained considerable clear serum. The heart and lungs were encased in a thick layer of purulent substance containing innumerable swine-plague bacteria. Lungs nearly collapsed; no hepatization.

Rabbit No. 130 died from the effect of an injection of 2 cc. of the sterilized emulsion. In cæcum several punctiform hemorrhages. Liver and kidneys were hyperæmic. The spleen was congested and somewhat enlarged. On the pleural side of the diaphragm were several punctiform hemorrhages; they were also present in the meninges of the spinal cord and in the subcutis over the cranium. No peritonitis or pleuritis. No swine-plague bacteria in cultures from the organs.

Rabbit No. 50 was killed February 25, 1892, nearly eight months after inoculation. On abdomen, at point of inoculation, there was an open abscess about 3 cm. in diameter. Its contents were of a grayish, pasty consistency. Otherwise the animal was in excellent condition. As the abscess was open no cultures were made from it.

Rabbit No. 53 was killed February 18, 1892, nearly eight months after inoculation. Rabbit in excellent condition. Beneath the skin over the abdomen there were two small abscesses; one about the size of a small walnut had ruptured, the other was about 3 cm. long and 1 cm. in diameter. It contained a pasty, yellowish, purulent substance. An examination showed only pus cells and degenerated cell substance. A tube of bouillon inoculated with a loop of the contents remained clear.

Rabbit No. 52 was killed February 12, 1892, seven and one-half months after inoculation. Rabbit in a well-nourished condition. At the point of inoculation on the side of the abdomen there was an abscess 8 by 3 cm., composed of a considerable number of small cysts which were found to communicate with each other. The contents consisted of a thick, pasty material composed of pus cells in a state of fatty degeneration. Lungs congested, hypostatic. Heart muscle pale, otherwise normal. A tube of bouillon inoculated with the blood remained clear. One inoculated from the contents of the abscess developed into a pure culture of the swine-plague germ, whose virulence was tested as follows:

February 18, a rabbit was inoculated subcutaneously with 0.2 cc. of the bouillon culture from rabbit No. 52. February 25, rabbit was found dead. Extensive purulent infiltration into the subcutis and inter-muscular tissue extending over an area 9 cm. in diameter, agglutinating the thigh to the abdominal wall. Cephalic lobes of both lungs in a state of gray hepatization containing small abscesses. A pure culture of the swine-plague germ was obtained from the spleen.

Rabbit No. 36 chloroformed February 15, 1892, eight and one-half months after its inoculation. Rabbit in a well-nourished condition. On the thorax there was an abscess about the size of a hen's egg which had ruptured. Surrounding this were several encysted abscesses varying from 0.5 to 2 cm. in diameter containing a thick pasty, purulent substance. These did not appear to have any communication with each other or with the ruptured abscess. Otherwise the rabbit was apparently normal. A tube of bouillon inoculated with a bit of the contents of one of the cysts developed into a pure culture of the swine-plague bacillus. On February 18, a rabbit was inoculated subcutaneously in the thigh with 0.2 cc. of this bouillon culture. It was found dead on the following morning. Innumerable swine-plague bacteria were found in stained cover-glass preparations from the various organs.

The remaining animals, which showed no external lesions or symptoms of disease after a few months, were used for other purposes.

The presence of swine-plague bacteria in the subcutaneous abscesses of rabbits Nos. 52 and 36 is a further illustration of the long period of time during which these organisms will remain alive in the lesions which they produce. Attention is called to the attenuated condition of the bacteria obtained from rabbit No. 52 and to the virulence of those isolated from a similar abscess in rabbit No. 36.

Although the rabbits killed for examination exhibited no lesions which presumably would not have been overcome had the animal been permitted to live, the fact remains that they were not made totally insusceptible to the action of the virulent bacteria. Here again is an illustration of the dangers of vaccination, for although the rabbits survived, they could not be considered as safe companions for other susceptible animals so long as they harbored the virus of the disease in abscesses which were liable to rupture at any time.

EXPERIMENTS WITH THE FILTRATE OF AGAR SUSPENSIONS.

The successful production of immunity in rabbits with sterilized turbid suspensions of swine-plague bacilli grown on the surface of nutrient agar made it desirable to test the immunizing properties of the bouillon in which these bacilli had been suspended for several hours and then removed by filtration. Such an experiment would inform us how far the bodies of the bacteria themselves, or any substance derived from them and passing promptly into solution in the suspending fluid, would be responsible for the immunizing action.

The suspensions from agar cultures were prepared as already described and placed in an incubator for twenty-four hours. They were then sterilized by heat as before and filtered through a Pasteur bougie, which removed all of the bacteria. The filtrate was perfectly clear and of a dark amber color.

Four rabbits were injected intravenously with the filtrate. Two of them received 3 cc. in three injections of 1 cc. each, and one received 3 cc. in two injections, and one received 4 cc. in three injections. They were subsequently inoculated with virulent swine-plague bacteria. They died in less than twenty-four hours with the control animals.

The injection of the filtrate was followed by a slight elevation of temperature and general depression, which was of much shorter duration than that produced by the suspension. A control experiment showed that similar symptoms were developed by the injection of an equal quantity of sterilized bouillon or normal salt solution.

EXPERIMENTS WITH DEFIBRINATED, STERILIZED BLOOD OF RABBITS AFFECTED WITH SWINE PLAGUE.

The culture of swine-plague bacteria used in the preceding experiments was also used here. It was fatal to rabbits twenty hours after inoculation.

The blood was obtained as follows: A rabbit was inoculated with a very small dose of swine-plague bacteria late in the afternoon, and on the following day it was watched very carefully, and just before death would have ensued the animal was etherized and the blood drawn by means of a slender sterilized glass canula inserted into the carotid artery. The blood was collected in large sterile tubes and defibrinated. It was then heated in a water bath for thirty minutes at a temperature of 58° C. Agar tubes were inoculated with the blood before and after the heating. The cultures made from the blood before it was heated showed the swine-plague bacteria to be present in large numbers. Those made after the heating remained invariably clear. The sterilized blood was kept at a temperature of about 8° C. until it was needed for use, when the desired quantity was heated up to the body temperature.

A.—EXPERIMENTS ON RABBITS.

In these experiments the sterilized blood was injected subcutaneously and into an ear vein. The number of injections made, quantity of sterilized blood used, and the results obtained after the test inoculation with the living bacteria are given in the following table:

XV.—*Table showing results of experiments with rabbits treated with sterilized blood of swine-plague rabbits.*

Date of blood injection.	Mar. 24.	Mar. 28.	Apr. 4.	Apr. 7.	Apr. 10.	Total.	Inoculated with swine-plague bacteria—	Results.
Subcutaneous.								
Rabbit—	cc.	cc.	cc.	cc.	cc.	cc.		
No. 441	1.5	1.5	3	3	3	12	Apr. 19, 1893	Recovered.
No. 446	1.5	1.5	3	3	3	12	....do	Do.
No. 443	1.5	1.5	1.5	1.5	....	6	....do	Do.
No. 442	....	1.5	1.5	3	....	6	....do	Died June 14, pleuritis and pericarditis.
No. 485(control)	....	....	....	....	....	....	....do	Died April 20, acute swine plague.
Intravenous.								
Rabbit—								
No. 439	1	1.5	1.5	....	....	4	Apr. 8, 1893	Recovered.
No. 440	1	1.5	1.5	....	....	4	....do	Do.
No. 444(control)	....	....	....	....	....	....	....do	Died Apr. 9.

The immediate effect produced by the injection of the sterilized blood was manifested by a considerable rise of temperature in about three hours after the injection. It soon subsided, however, and no other symptoms were detected. This was true of the rabbits that were injected both intravenously and beneath the skin. The temperature of the treated rabbits was normal at the time the control animals died. For several weeks these rabbits appeared entirely well. It was then observed that subcutaneous abscesses were forming on different parts of the body. Nearly two months after the check died rabbit No. 442 was found dead. The post-mortem notes of this animal are as follows:

June 14, rabbit No. 442 found dead. It was somewhat emaciated. The abdominal organs were not appreciably diseased. The pleural cavity contained a large quantity of serum, and the pleura (visceral and parietal) was covered with a thick exudate, the deeper layer of which was firmly adherent to it. The heart was incased in a rather thick membranous exudate. Heart muscle pale; blood dark but not clotted. Cover-glass preparations from the exudate contained a considerable number of polar-stained swine-plague bacteria. A tube of bouillon inoculated with the exudate developed a pure culture of swine-plague bacteria.

June 15, rabbit No. 443 was chloroformed. The post-mortem examination showed a slight thickening of the subcutaneous tissue at the point of inoculation. The thoracic and abdominal organs were apparently in a normal condition. No swine-plague bacteria were found in the organs or in the subcutaneous tissue at the point of inoculation.

June 30, 1893. The abscesses beneath the skin of rabbits Nos. 441, 446, 439, and 440 were found to vary from 1 to 4 cm. in diameter. None of them were located at the point of inoculation. Their distribution is best illustrated by referring to No. 439. This rabbit had an abscess, which had ruptured, over the sternum, an abscess as large as an English walnut in the subcutaneous tissue on the right side of the thorax, one beneath the superficial muscle over the right thigh, and one small abscess beneath the skin in the right popliteal space. The abscesses were opened in rabbits Nos. 439 and 448 and found to contain a thick, somewhat viscid, grayish, purulent substance. Tubes of bouillon inoculated from this purulent substance developed into pure cultures of swine-plague bacteria. This localization of the bacteria did not appear to affect the general health of the rabbits. After the abscesses were opened they healed with one exception (one abscess in rabbit No. 439), and the rabbits appeared to be perfectly well. In order to determine whether or not the abscesses were confined to the subcutaneous tissue, these rabbits (Nos. 439 and 440) were chloroformed July 14, and thoroughly examined. No. 439 showed no lesions, but in No. 440 two of the axillary glands were enlarged and found to contain foci of suppuration, and beneath the peritoneum in the lumbar region there was an abscess containing broken-down pus cells and swine-plague bacteria, as determined by microscopical examination. The kidneys were small and firm but free from bacteria, as indicated by cultures. Nos. 441 and 446 were inoculated twice subsequently with a culture of the same virulent swine-plague bacteria without manifesting any general disturbances. For their further history and use see p. 74.

For their further history and use see p. 74.

B.—EXPERIMENTS ON GUINEA-PIGS.

A single experiment was made on guinea-pigs with the sterilized blood from the swine-plague rabbits. Two guinea-pigs, Nos. 374 and 377, received subcutaneously 5.25 cc. each of the sterilized blood in 3 doses (March 28, April 4, and April 7). April 18 they were inoculated subcutaneously, together with two control guinea-pigs, with 0.01 cc. each of a fresh bouillon culture of swine-plague bacteria. The check died in six days. The treated animals remained apparently perfectly well. They were subsequently used for other purposes, but on post-mortem examination they did not reveal the existence of any lesions that could be attributed to the swine-plague bacteria previously inoculated.

GUINEA-PIGS MADE INSUSCEPTIBLE TO SWINE-PLAGUE BACTERIA OFFER NO RESISTANCE TO HOG-CHOLERA BACTERIA, AND VICE VERSA.

The difference that exists between the products of hog-cholera and swine-plague bacteria, as shown by the foregoing experiment, is further illustrated by the fact that guinea-pigs that are immune to one disease offer no resistance to the other. From the differences already pointed out between these two species of bacteria this condition would be expected, and consequently only one experiment has been made with each.

Guinea-pigs Nos. 1 and 2, that had been immunized against hog cholera by subcutaneous injections of sterilized bouillon cultures, were inoculated with 0.01 cc. of a bouillon culture of virulent swine-plague bacteria. They died on the third day with the control animal. One of these guinea-pigs (No. 2) had been twice inoculated with hog-cholera bacteria.

Guinea-pigs Nos. 374 and 377 were made resistant to the fatal effect of swine-plague bacteria by subcutaneous injections of sterilized defibrinated blood from swine-plague rabbits. They were subsequently inoculated with 0.1 cc. of a bouillon culture of virulent hog-cholera bacteria. They lived one and two days respectively longer than the control, which died in an unusually short time. The lesions found upon examination were those characteristic of acute hog cholera. The variations found to exist in the length of time required for 0.1 cc. of a bouillon culture of hog-cholera bacteria to kill guinea-pigs render the short time these animals lived after the control died of no significance, especially as there were no variations in the lesions to indicate increased resistance.

PROTECTIVE ACTION OF BLOOD SERUM FROM IMMUNE RABBITS.

According to Metchnikoff a comparatively small quantity of blood serum from immunized animals is sufficient to induce immunity in rabbits. It will be remembered that while he asserts this for hog-cholera bacilli he was actually working with swine-plague bacteria. As small a quantity as 0.5 cc. of serum from immunized rabbits injected subcutaneously was sufficient to protect a rabbit from a quantity of virulent blood sufficient to kill a control rabbit. Injection of serum into the circulation directly was much less efficacious.

The following experiment fully confirms Metchnikoff's work if, as has been stated above, we substitute the words swine plague for hog cholera in his article. The serum was obtained from rabbits Nos. 441 and 446 made immune with sterilized defibrinated blood of rabbits affected with swine plague and referred to on page 72.

These rabbits were reinoculated with swine-plague bacteria as follows:

June 24, 1893. Nos. 441, 446, and a control receive subcutaneously 0.12 cc. of a bouillon culture three days old.

June 25. Control dead this morning. The others apparently unaffected.

December 5, 1893. Nos. 441, 446, and a control received subcutaneously 0.12 cc. of a bouillon culture (the same stock culture) twenty-four hours old.

December 6. Control dead.

December 7. No. 441 has a temperature of 102° F.; No. 446, 101° F.

December 12. No. 441 has a temperature of 102° F.; No. 446, 101.4° F.

No. 446 was etherized and bled from a carotid January 4, 1894, thirty days after the last inoculation. All the blood obtainable was collected and the animal allowed to die while anæsthetized. It was very thin at this time. In the lungs there were in all five foci of disease in which the lung tissue was converted into a necrotic mass containing large numbers of pus cells and many polar-stained bacteria. In the bronchi and trachea a considerable quantity of muco-purulent material containing not the polar-stained bacteria but small cocci deeply stained, in masses and mainly within pus cells. The lung lesions where they reached the pleura had produced an exudate attaching the lungs to the chest wall by means of still soft, easily broken adhesions.

The cause of these lung lesions was not definitely determinable. At first thought the subcutaneous inoculation thirty days ago appeared responsible, but the location of the disease did not harmonize with this supposition. Usually the pleura is involved first and very extensively, and the lungs only secondarily by contiguity. Furthermore, the bacteria present appeared slightly larger than those injected and were very much attenuated, so that it required a comparatively large intravenous dose to destroy a rabbit. Parallel bouillon cultures of this bacillus differed quite markedly in appearance from those of the swine-plague bacillus originally injected.

The more probable supposition appeared to be that this rabbit was infected through the air passages with the bacillus of rabbit septicæmia* or influenza recently investigated by one of us in this laboratory, which bacillus resembled the one from rabbit 446 closely, both in cultures and in pathogenic activity.

Rabbit No. 441 was bled to death under ether January 8. This animal was in good condition and no internal lesions were found.

---

*Moore and Kilborne: An outbreak of rabbit septicæmia, with observations on the nature of the disease and its specific organism. American Veterinary Review. 1893, Vol. xvii, p. 285.

Cultures from spleen and liver remained sterile. The serum was used as indicated in the subjoined table:

XVI.—*Table showing results of protective action of blood serum.*

Blood from .... No. 446.	No. 441.							Inoculated with 0.06cc. virulent bouillon culture.	Result.
When drawn .... Jan. 4.	January 8.								
Date of injection of serum .... Jan. 5.	Jan. 9.	Jan. 10.	Jan. 11.	Jan. 12.	Jan. 13.	Jan. 15.	Total.		
Rabbit— *cc.*	*cc.*	*cc.*	*cc.*	*cc.*	*cc.*	*cc.*	*cc.*		
No. 120................	0.9	1	1	.....	0.5	0.5	3.9	Jan. 19....	Recovered.
No. 121............. 0.5	0.9	1	1	1	0.5	0.5	5.4	Jan. 19....	Dies Feb. 5.
No. 130................				1	1.5	..... 0.4	2.9	Jan. 19....	Dies Feb. 12.
No. 129 (control) ...								Jan. 19....	Dead in 20 hours.
Guinea-pig—									
No. 39................				1.3	1.0	0.75	3.05	Jan. 19....	Recovered.
No. 40................				1.3	1.0	0.75	3.05	Jan. 19....	Dead Jan. 30.
No. 41 (control)								Jan. 19....	Dead Feb. 1.

The subsequent history of the three treated rabbits and the guinea-pigs is briefly told.

No. 121. January 22, three days after the test inoculation, the temperature was 103.3. A local swelling was noticed, about one-half inch in diameter and one-eighth inch thick.

January 24.—Temperature, 102.5. Local swelling larger, 1 inch across at base, one-half inch thick.

January 29.—Lesion as before. Up to to-day this rabbit has been eating, moving about, and playful.

February 5.—Rabbit dies very unexpectedly to-day. It was still playful this morning. The autopsy showed that the abscess at the place of inoculation had gradually destroyed the abdominal wall over an area one-fourth of an inch in diameter. The contents of the abscess, protected only by the peritoneum, projected into the abdominal cavity as a conical mass. The swine-plague bacteria had penetrated this delicate barrier and produced a fatal peritonitis. The viscid exudate contains immense numbers of swine-plague bacteria and a small number of leucocytes. Phagocytosis absent. The bacteria from this partly immune rabbit had not been attenuated. An inoculated rabbit died within 20 hours.

No. 120. On January 22, swelling 1 inch at base, one-half inch thick. Temperature, 102.2.

January 24.—Swelling much larger. Temperature, 102.9.

January 24. Swelling as large as a hen's egg. Up to date this rabbit has been more quiet than No. 121.

March 1. Rabbit now fully recovered. Weight has fallen from 2.5 to 2 pounds. On March 28, a slight gain in weight.

May 15. Has steadily increased in weight.

No. 130. On January 22, temperature 103.4. Swelling like that of No. 120.

January 24. Swelling larger. Temperature, 104.2.

January 29. Swelling as before. Skin breaking. This animal appears quiet but not sick.

February 12. Rabbit dies quite suddenly to-day, though apparently well for a week. Local lesion almost healed. No lesions referable to swine plague excepting perhaps a slightly roughened, opaque condition of the serous covering of cæcum.

The experiment with the guinea-pigs did not give any clear results, as is shown in the table. No. 40 died January 30 with exudative pleuritis and markedly hyperæmic lungs. The control died two days later with same lesions. In both the exudate contained immense numbers of swine-plague bacteria. The third guinea pig (No. 39) recovered, but did not appear to thrive. It was chloroformed two months after the test inocculation, but no lesions were discovered. This equivocal outcome thus opens the question whether perhaps the immune rabbit's serum is less efficacious when used upon some other species, such as guinea-pigs, than when used on the same species; for the results obtained with rabbit's serum on rabbits are strikingly positive.

DOES THE BLOOD SERUM OF IMMUNE RABBITS POSSESS ANY BAC-TERICIDE OR ANTITOXIC POWER?

The following experiment was tried with the blood serum of No. 441: To a given quantity, 0.6 cc., was added a small quantity of a bouillon culture twenty-four hours old, equivalent in amount to one-tenth of the blood serum. This mixture was injected subcutaneously. The experiment is tabulated below:

XVII.—*Table showing results of injection of mixture of blood serum and bouillon culture.*

Rabbit—	Serum mixture contains—		Age of mixture when injected subcutaneously.	Result.
	Bouillon culture swine plague.	Serum.		
	cc.	cc.		
No. 474 (5½ pounds).............	.066	0.6	10 minutes......	Dies in 24 hours.
No. 119 (3 pounds).......	.066	0.6	5 hours	Dies in 60 hours.
No. 118 (3 pounds, control).....	.066	0.6 sterile bouillon.	....	Dies within 20 hours.

In the rabbit receiving the serum mixture five hours old the disease was markedly retarded. On the following day, when the other rabbits were dead, it was quite active and ate its food as usual. On the second day the temperature had risen to 107.6° F. On the morning of the third day it was found dead. The bacteria had not been modified in this rabbit, for a minute, subcutaneous dose from a culture was fatal to a rabbit in 20 hours.

CONCLUSIONS.

The preceding observations and experiments have led to a few well-defined results, which we wish to summarize briefly. The hidden, underlying, vital processes, according to which these results may be interpreted, are either matters of controversy or else wholly intangible as yet. We content ourselves, therefore, in pointing out the more gross, incontrovertible facts elicited, leaving the discussion of subtler problems and the citation of the mainly controversial literature for the present untouched.

Immunity defined as a resistance to the invasion and multiplication of pathogenic bacteria within the body is, when compared with susceptibility, a condition of degree rather than of kind. There are various degrees of immunity producible by one or the other of the methods employed in these experiments, and in speaking of immunity experimentally produced it is necessary to state precisely the conditions under which such immunity is manifested. Thus a rabbit may be treated so as to resist successfully a fatal injection of swine-plague bacteria into the subcutaneous tissue. The same injection into the abdominal cavity or into the circulatory system may result in speedy death. Continued preventive treatment may indeed finally result in resistance to even these fatal doses. Our experiments have not been pushed to these extremes, and we are therefore unable to state whether, for instance, rabbits may be brought to this high level of immunity without a permanent injury to the organs and tissues, which injury manifests itself in the lower stages of the immunizing process by more or less prolonged emaciation. In the work before us the highest degree of immunity sought for in rabbits and guinea-pigs towards the bacteria of swine plague and hog cholera was a permanent resistance to those minimum doses of culture material which result fatally when injected under the skin of untreated animals.

The various processes we have employed in the production of immunity may be classed as follows:

(a) The use of attenuated cultures of living bacteria.

(b) The use of sterilized bouillon cultures.

(c) The use of sterilized bouillon suspensions of bacteria from agar cultures.

(d) The use of sterilized blood from animals in the last stages of the fatal inoculation disease.

(e) The use of blood serum from susceptible animals in which a certain degree of immunity had been produced by one or the other of the foregoing methods.

These five processes may be ranged under three main heads:

A (including a). The use of living bacteria.

B (including b, c, d). The use of the chemical products of bacteria.

C (including d, e). The use of certain still undefined substances in the blood, variously called antitoxins, alexins, etc. *

It may be that all of the methods given will be ranged, after more exhaustive investigations, under bacterial products, a view even now maintained by Büchner.† The relative efficiency of these processes is not precisely the same for both diseases experimented with, and for this reason, as well as for the sake of greater clearness, the result obtained with each species is best summarized by itself and comparisons made subsequently.

HOG CHOLERA.

If we compare the protective action on rabbits we shall find that of all the methods, the use of living, attenuated cultures was the only one which led to success. Even with this method the failures were many because the vaccinal doses were often too great or the culture of too high a degree of virulence. Rabbits are so susceptible to hog-cholera

---

* d is included under B and C, because it is proper to assume that in the blood of sick animals we have antitoxins as well as bacterial products present.

† Münchener med. Wochenschrift, 1893, Nos. 24, 25.

bacteria that their immunization requires, as has been shown in the text, not less than two preventive inoculations at an interval of one or more months. It is not to be denied that immunity may be produced with the use of sterilized agar suspensions, sterilized blood of diseased rabbits or blood serum from immune rabbits, but these methods as applied by us made no impression on rabbits whatever. When applied in the same way in swine plague they produced a high degree of resistance.

The less susceptible guinea pig reacted somewhat differently. Immunity was brought about with sterilized bouillon cultures but with varying results. Living attenuated cultures were not tried. Sterilized agar suspensions produced a slight retardation of the disease while sterilized blood from diseased guinea-pigs failed to produce any impression. A slight retardation of the disease was also produced by the blood serum of immunized guinea-pigs but no retardation was noticed when the blood serum from immunized rabbits was used.

SWINE PLAGUE.

A greater or less degree of immunity was produced in rabbits by sterilized bouillon cultures, sterilized agar suspensions, sterilized blood from infected rabbits, and blood serum from immunized rabbits. Experiments on guinea-pigs were restricted to two methods. The sterilized blood of diseased rabbits was capable of producing immunity while the blood serum of immune rabbits produced rather equivocal results, as Table XVI shows.

In the following table those experiments are marked with the sign + in which more or less immunity was produced, and those with the sign — in which no appreciable resistance was noticed. Even those in which only a distinct retardation of the fatal issue was produced are considered positive, for it is fair to assume that such retardation would have become immunity if the treatment had been pushed a little further:

Immunizing processes employed.	Rabbits.		Guinea-pigs.	
	Hog cholera.	Swine plague.	Hog cholera.	Swine plague.
Repeated injection of living attenuated cultures.	·			
Sterilized bouillon cultures....	—	+ (16 cc. in 5 injections, ear vein.)	+ (10 cc. in 4 injections, 8 cc. in 8 injections.)	
Sterilized suspensions of agar cultures.	—	+ (7-8 cc. in 4 injections, abdomen.)*	Very slight effect.	
Sterilized blood of rabbits in last stage of disease.	—	+ (6 cc. in 4 injections, subcutis.)	—	+ (5 cc. in 3 injections, subcutis.)
Sterilized blood of guinea pigs in last stages of disease.			—	
Blood serum from immunized rabbits.	—	+ (4 cc. in 5 injections, subcutis.)	Doubtful........	Doubtful.
Blood serum from immunized guinea-pigs.			+ (5 cc. in 5 injections, produced retardation).	

* Immunity produced in some cases with smaller doses. See Table XIV.

The experiments on swine with suspensions of agar cultures of hog-cholera bacilli demonstrate that immunity toward a fatal intravenous dose may be produced. Unfortunately no opportunity was offered to test the behavior of pigs protected in this way when exposed to the natural disease on the farm. The cost of preparing the sterilized suspensions would be a decided objection to their employment on a large scale.

We are well aware of the fact that only a beginning has been made by us in the study of experimental immunity with reference to these two important animal disease germs. A continued investigation of the collateral problems which have presented themselves in the course of the work is highly desirable, especially in the direction of seeking an answer to the question, why some species of animals are very susceptible, others only partially so, to the same pathogenic bacteria.

The definite progress made in this work is illustrated by the following important determinations:

1. It is possible to produce immunity toward hog-cholera and swine-plague bacteria in the very susceptible rabbit and the less susceptible guinea-pig. In the rabbit the only promising method of immunization toward hog cholera is the use of gradually augmented doses of attenuated cultures.

2. Immunization toward swine-plague bacteria is produced artificially with much greater ease than toward hog-cholera bacteria.

3. The blood serum of animals protected against hog cholera and swine plague is almost as efficacious in producing immunity soon after treatment as the bacterial products obtained from cultures.

4. Different degrees of immunity in both hog cholera and swine plague lead to different forms of the inoculation disease. The greater the immunity short of complete protection the more prolonged and chronic the disease, induced subsequently by inoculation.

5. Pathogenic bacteria may remain in the organs of inoculated animals some time after apparently full recovery. Their presence may or may not be associated with lesions recognizable by the naked eye.

6. The toxicity of sterilized cultures appears to be directly proportional to the number of bacteria in the injected fluid.

7. The results of Selander and Metchnikoff in the immunization of small animals were obtained with swine-plague and not with hog-cholera bacilli.

ON THE VARIABILITY OF INFECTIOUS DISEASES AS ILLUSTRATED BY HOG CHOLERA AND SWINE PLAGUE.

By THEOBALD SMITH and VERANUS A. MOORE.

The ideas incorporated in this article date back a number of years and owe their origin to the work on swine diseases carried on since 1885.

The first intimation of any variation in the virulence of bacteria from swine diseases was obtained by the isolation of swine-plague bacteria from different outbreaks.* Some of these varieties produced in rabbits after subcutaneous inoculation an exudative peritonitis fatal within a week. Others produced a septicæmia fatal within twenty-four hours.

In 1889 one of us pointed out the difference in the lesions produced in rabbits by the inoculation of two distinct varieties of hog cholera.†

Out of these observations grew the question whether these bacteria, morphologically and biologically identical, must be considered as different because of the different degrees of pathogenic power manifested by them. A long series of inoculations into rabbits made in 1889 with two hog-cholera bacilli of different virulence showed that the diseases induced by them could be made the same either by reducing the virulence of one of the varieties or increasing the resistance of the inoculated animal. The same was demonstrated by a preliminary experiment with virulent swine-plague bacteria in 1889.‡

The continuation of the investigations on preventive inoculation has enabled us to collect a considerable amount of information illustrative of variations in the gross manifestation of disease which may be brought about experimentally. This information we deem of great importance toward the proper understanding of infectious diseases as a whole, and hence present it here in as compact a form as possible.

---

* See Report on Swine Plague (1891) for a description of these bacteria.

† Bacillus α and Bacillus ζ, pages 9 and 13. See also the New York Medical Journal 1890—ii, p. 485.

‡ Report on Swine Plague, p. 148.

22275—No. 6——6

SWINE PLAGUE.

Among the forms of disease which we have observed after the subcutaneous inoculation of rabbits with swine-plague bacteria from dif
ferent sources are the following:

1. Septicæmia.
2. Peritonitis.
3. Pleuritis (usually with pericarditis).
4. Pleuritis (usually with pericarditis) and peritonitis.
5. Local lesion only.

In the septicæmia death ensues within eighteen to twenty-four hours.
The local lesion produced at the seat of inoculation is slight. Bacteria
are abundant in the parenchyma (blood vessels) of the various organs.
In the form characterized by peritonitis death ensues in three to seven
days. The local lesion, which in all these forms of disease increases in
extent with the prolongation of the life of the animal, is here characterized by more or less suppurative infiltration of the skin and the subcutis. The peritonitis in its earlier stages is characterized by punctiform hemorrhages on the cæcum and an exudate varying in character,
being fibrinous or cellular. It always contains immense numbers of
bacteria. When pleuritis is also present the exudate usually involves
the pericardium as well. It varies in amount according to the duration
of the disease and is essentially the same as the peritoneal exudate.

The form characterized by pleuritis and pericarditis without peritonitis is interesting in so far as the seat of inoculation does not
explain the localization, for, in every case, the inoculation was made
in the region of the abdomen. The lungs may become hepatized secondarily through invasion from the pleura if the animal lives long
enough.

Lastly, the form of disease in which the only localization is a very
extensive suppurative infiltration associated with hemorrhage and
œdema of the subcutaneous tissue is not common.

It should be stated that the cultures from the same outbreak continued to produce the same form of disease in rabbits until modified by
age. The maintenance of a certain uniform virulence for years is well
exemplified by a variety isolated in the summer of 1890.* This variety
was fatal to rabbits within twenty hours when first isolated, and this
degree of virulence has maintained itself up to the present, a period of
nearly four years.

Inasmuch as numerous illustrations of these forms of swine-plague
in rabbits may be found in the report on swine-plague and in the
reports of the Bureau of Animal Industry since 1886, we refrain here
from citing any cases.

* Bulletin on Swine Plague, p. 57.

MODIFICATIONS OF THE SEPTICÆMIA TYPE BY INCREASING THE RESISTANCE OF RABBITS.

By the various processes recorded in the preceding article which increase the resistance of rabbits we have been able to produce nearly all the pathological variations which follow the inoculation of natural races of swine-plague bacteria as isolated from outbreaks. This modification of the septicæmia type is not fortuitous, for among the large number of rabbits inoculated during the past three and one-half years with the culture employed none have survived twenty to twenty-four hours. Whenever the course of the inoculation disease in rabbits departed from this rapidly fatal type it was due to some preliminary treatment of the rabbit.

The degree of resistance determined quite regularly though not invariably the form of the disease. This degree was measured by the relative quantity of the protective material (sterilized cultures, sterilized blood, and blood serum) injected. The grades of disease induced range themselves in the following order:

1. No resistance—acute septicæmia.
2. Slight resistance—peritonitis.
3. Increased resistance—pleuritis and pericarditis with or without secondary pneumonia.
4. Higher degree of resistance—pleuritis and peritonitis.
5. Still greater resistance—irregular lesions in the form of abscesses, subcutaneous and subperitoneal.
6. Nearly complete immunity. Very slight reaction at the point of inoculation.

Most of the cases cited below as illustrating these modified forms of the septicæmia type belong to the series of immunizing experiments of the preceding article. To this the reader is referred for additional illustrations.

First degree of resistance—peritonitis.—Rabbit No. 12 received 7 cc. of bouillon culture of swine-plague bacteria sterilized by heat. Subsequently with a control rabbit it was inoculated with a minute dose of swine-plague bacteria under the skin. The control died within eighteen hours, the treated rabbit in three days. The macroscopic changes were limited to the point of inoculation and the peritoneum. At the former there was a purulent infiltration of the subcutis, 1.5 cm. in diameter, with dilatation of surrounding blood vessels. The peritonitis was characterized by an exudate of a slightly viscid character covering liver, spleen, and cæcum, and made up of fibrin, leucocytes, and immense numbers of bacteria.

Second degree of resistance—pleuritis and pericarditis.—Rabbit No. 38 * was treated before inoculation with 4.5 cc. of a sterilized suspension of agar cultures of swine-plague bacteria in 3 doses. Together with a control rabbit, it received under the skin the equivalent of 0.001 cc. of a fresh bouillon culture of swine-plague bacteria. The

---

* *See* Bulletin on Swine Plague, p. 148, for the entire experiment tabulated.

control died in twenty hours. The treated rabbit died six days after
inoculation. At the point of inoculation there was a purulent infiltration of the subcutis 3 cm. in diameter. The abdomen and abdominal
viscera were free from macroscopic changes. In the thorax, the pleural
cavity was lined with a grayish, friable exudate consisting of round
cells and bacteria. Lungs hyperæmic and only partly collapsed.
Pericardium also covered with a slight exudate.

   Third degree of resistance—pleuritis (pericarditis) and peritonitis.—
Rabbit No. 15 received in the ear vein 12 cc. of a sterilized bouillon culture of swine-plague bacteria. It was inoculated subcutaneously with
virulent swine-plague bacteria May 26, and died June 3, eight days
later. The control rabbit died within eighteen hours. The following
changes were observed:

   A purulent infiltration into the subcutaneous tissue at the point of inoculation
extending over an area 6 cm. in diameter. The superficial layer of the subjacent
muscle discolored. Surrounding the area of infiltration the blood vessels were
injected. The cæcum and liver were covered with a very thin grayish exudate, which
also appeared on and between the coils of the intestine. Spleen not enlarged.

   The right lung and chest wall covered with a thin grayish exudate. In the cephalic
lobe, two small areas of consolidation; principal lobe hyperæmic. The left pleural
cavity lined with a quite thick membranous exudate, which covered the entire surface of the lung. On the dorsal surface of the principal lobe a mass of lung tissue
2 cm. in diameter, firm and of a yellowish-gray color. The remaining portion of the
principal lobe hyperæmic; cephalic lobe in state of collapse.

   Pericardium covered with a thin cellular exudate.

   Higher degrees of resistance.—None of the treated animals which have
come under our observation, have succumbed to a mere extension of
the lesion produced at the point of inoculation as is occasionally observed
after inoculation with certain varieties of swine-plague bacteria found
in nature. There have been noticed, however, certain peculiar localizations resembling those produced in the subcutis after inoculation, and
in a few cases the local lesion persisted a considerable length of time.
It was quite severe in all fatal cases in which the disease was prolonged
several weeks after inoculation, although the real cause of death was
due in all such cases to localizations on one or more of the serous membranes. The peculiar forms of disease may be grouped as follows:

   (*a*) *Persistence of local lesion.*—Rabbit No. 50 received in the abdominal cavity 3.5
cc. of the sterilized suspension of agar cultures in 3 doses. It was subsequently
inoculated beneath the skin with 0.001 cc. of a bouillon culture of swine-plague bacteria which produced a large local swelling. On February 25, 1892, nearly eight
months after its inoculation, it was chloroformed. The only lesion found was in the
subcutaneous tissue. At the point of inoculation the skin was sloughed over an area
3 cm. in diameter. This denuded surface was covered with a thick scab. The subcutis beneath the scab and surrounding the ulcer was infiltrated with pus. A stained
cover-glass preparation showed swine-plague bacteria. No other lesions were found.

   (*b*) *Sub-peritoneal abscess.*—Rabbit No. 16 was injected intravenously with 16 cc. of
sterilized bouillon cultures of swine-plague bacteria. After some days it was inoculated beneath the skin with 0.001 cc. of a fresh bouillon culture of virulent swine-

plague bacteria. The control rabbit died within twenty hours. Rabbit No. 16 showed no ill effect from the inoculation for several months when it was noticed that it was becoming emaciated. It died June 11, 1892, one year and six days after its inoculation with an enormous subperitoneal tumor, which is described on page 67.

(c) *Multiple abscesses under the skin.*—Rabbit No. 439 received into the ear vein in 3 injections 4 cc. of sterilized blood from a swine-plague rabbit. Later it was inoculated subcutaneously with 0.001 cc. of a bouillon culture of virulent swine-plague bacteria. The control rabbit died within twenty hours. Two months after the inoculation it was noticed that this rabbit with others was suffering from a large number of subcutaneous abscesses, described on page 72.

HOG CHOLERA.

The lesions produced in rabbits after subcutaneous injection of the most virulent variety of the hog-cholera bacillus (the one most frequently encountered) have been given on page 10. The lesions produced by the less virulent varieties may either resemble these on intravenous inoculation or else they may be quite different both after subcutaneous and intravenous inoculation. That variety of the hog-cholera bacillus described as ζ on page 13 presents some striking differences, to which reference has already been made by various illustrative cases. In the illustrations given below the macroscopic changes in those animals which succumbed to a more prolonged chronic illness were concentrated in the follicular-apparatus of the intestines and the lungs.

Modified disease produced in a rabbit with hog-cholera bacilli attenuated by heat.—The culture used had been exposed for thirty-eight days to 43.5° C. to 44° C., according to the procedure described on page 42. The rabbit received subcutaneously 0.12 cc. of a fresh bouillon culture twenty-four hours old. It died in twenty days. The temperature during the disease remained rather low, fluctuating between 103° and 104° F. On the fourteenth day it began to sneeze and cough and a few days later a muco-purulent discharge from the nose appeared. On the twentieth day it was unable to get up and was chloroformed. The lesions were in brief as follows:

At the point of inoculation a small abscess; spleen moderately enlarged; liver free from necroses; the bunch of lymph glands at root of mesentery as large as a horse chestnut and mottled with whitish foci; the Peyer's patches at ileo-cæcal valve are thickened; the follicles very large, whitish; the overlying mucosa ulcerated; the lungs were also involved; the entire right ventral and a portion of the right cephalic lobe involved in broncho-pneumonia; the left principal lobe is solid, dark red on section; the air passages contain a thick muco-pus.

Portions of an ulcerated Peyer's patch, of the enlarged mesenteric glands, and of the lungs hardened in alcohol were submitted to microscopic examination. The follicles and the follicular tissue of the Peyer's patch owe their swollen condition to leucocytic infiltration in the depths of which a few clumps of hog-cholera bacilli are brought out by methylene blue. In the gland the whitish foci were found to be collection of leucocytes, largely broken down, in which are embedded clumps of bacilli. The pneumonia was characterized by dense cell infiltration of alveoli and small air tubes with desquamation of cells in the alveoli around the broncho-pneumonic foci.

A rabbit was inoculated subcutaneously with 0.5 cc. of a culture of the hog-cholera bacillus exposed to 43.5° to 44° C. for one hundred and ninety-five days. Rabbits inoculated at the same time with smaller doses survived, while this one died, very much emaciated, in thirty-seven days. The local lesion was encysted, and the only other changes consisted of twelve infiltrated follicles in the appendix and about six similarly affected follicles in the Peyer's patches in ileum and cæcum, near the valve. The mucosa overlying the cæcal patch shows two excavated ulcers with irregular edges.

A rabbit was inoculated by injecting 0.12 cc. of peptone bouillon culture into an ear vein. The bacilli used were from the culture described as β (page 10), and at the time of injection nearly seven years old. The attenuation was thus due to prolonged cultivation. The rabbit succumbed in four days. Besides the usual parenchymatous changes, there had developed a pneumonia, involving the right ventral and azygos lobe in hepatization. The trachea and bronchi contained much catarrhal exudate.

The lesions produced by the inoculation of virulent material into rabbits, in which a partial immunity has been induced are similar to those produced by attenuated cultures in rabbits having no artificial protection. This is well illustrated by the following case:

Rabbit No. 13. June 12, 1889, receives subcutaneously 0.05 cc. bouillon culture of virulent hog-cholera bacilli exposed to 43.5° to 44° C. for ninety-eight days. The temperature rose to 106.8° F. four days after inoculation and fell to normal in two weeks.

July 16. Subcutaneous inoculation of 0.05 cc. unattenuated culture.

October 1. Subcutaneous inoculation with a particle of spleen pulp from a rabbit.

December 14. Found dead to-day. The lesions are very interesting as compared with acute hog cholera.

Intestines.—Much mucus present in small and large intestines. Peyer's patch in cæcum near valve infiltrated and on the mucous surface converted into a yellowish brown slough. In the cæcum itself along the mucous folds are groups of slightly depressed firm sloughs of the mucosa. Similar ulcers in upper portion of colon. In the appendix all the follicles are whitish and more or less swollen, some 4 to 5 mm. in diameter and projecting into lumen of tube. Mucosa not necrosed.

Spleen large, dark, and soft. Liver dotted with a considerable number of grayish spots. The presence of coccidia makes their nature doubtful. On the surface of both kidneys there are 50 to 75 firm nodules, variable in size, tho largest being 4 mm. in diameter·and projecting hemispherically above the surface. They dip down into cortex as.elongated masses of a caseous appearance and firm consistency.

The lungs remain fully expanded on opening thorax. With the exception of a narrow border on cephalic lobes they are completely hepatized. The surface, of a bright red color, is beset with pale yellowish irregular spots 1 to 4 mm. in diameter. These spots are not well defined, but surrounded by a hazy nebulous border. They correspond to nodules in the lung tissue which are of a cheesy consistency and crumble on pressure.

In sections of hardened tissue the lung is found containing necrotic foci surrounded by a zone of alveoli filled with round cells and more or less new connective tissue.

The hog-cholera bacillus was still present in the spleen. A colony from an agar culture produced the characteristic acute fatal disease in rabbits. Any attenuation of the bacillus not noticeable.

The localization of hog-cholera bacilli, injected into the circulation, in the brain substance, and the production of a focus of necrosis and suppuration, is illustrated by the rabbit referred to on page 43 of this bulletin. In this case macroscopic changes in other organs were not detected.

A FORM OF PSEUDO-TUBERCULOSIS PRODUCED IN RABBITS AND GUINEA-PIGS AS A RESULT OF INCREASED RESISTANCE OF THE ANIMALS OR ATTENUATION OF THE VIRUS.

A.—GUINEA-PIGS.

This form of inoculation disease was first noticed in guinea-pigs in 1890. The resistance of the guinea-pigs had been increased by various methods. Some of them survived the inoculation of the usually fatal dose of hog-cholera bacilli. Others succumbed to a prolonged disease, during which the pseudo-tubercles to be described were formed.

These tubercles situated, as a rule, in the subserous tissue of the abdomen and under the pulmonary pleura, are small, slightly convex bodies of a whitish or neutral gray color. In form they are roundish or slightly elongated. In size they vary from barely recognizable dots to bodies 2 mm. in diameter. They are firm in consistency and give the sensation of hard, granular bodies. In appearance they are strikingly like the true tubercles when the latter are situated on the serous membranes. In fact when they were first noticed it was thought that the guinea-pig had been inoculated accidentally with tuberculous material.

In sections, cut by the paraffin method, these bodies were found to consist of aggregations of round cells of which the central portion was largely disintegrated. No limiting membrane, stroma, or giant cells have been found. When crushed with forceps and examined under the microscope in a fresh condition cells more or less degenerated are the only tissue elements observable. In properly stained cover-glass preparations made from the crushed bodies hog-cholera bacteria were found in considerable numbers. Similar preparations from the blood, spleen, and liver of the same animal failed to show them.

These pseudo-tubercles were almost always present in guinea-pigs which lived four or more days longer than the control animals. In a similar series of experiments made by Dr. E. A. de Schweinitz in 1890 with substances isolated from pure cultures of hog-cholera bacteria the pseudo-tubercles were found when an increased resistance equivalent to only one or two days had been induced.*

In the experiments to test the immunizing efficacy of sterilized cultures of hog-cholera bacteria (p. 46) thirteen guinea-pigs died in four or more days after the control animals. In nine of these the pseudo-tubercles were found.† It is of interest to note that in a few guinea-pigs

---

*Medical News, October 4, 1890.
† For the history of the four other guinea-pigs (Nos. 477, 19, 478, and 438), see p. 47.

that were killed for examination several months after inoculation the tubercle-like bodies were not found.

The distribution* of the pseudo-tubercles in the body of the guinea-pigs varied considerably in the different animals. They were either quite generally distributed beneath the serous membranes or confined within circumscribed areas, where they came into actual contact one with another or were separated by a distance of from 0.5 to 3 mm. The localities in which they have been found, beginning with the most common, are as follows:

1. Beneath the peritoneum, more especially along one side of the spinal column or else aggregated in limited areas from 3 to 5 cm. in diameter. When separated by only a short distance the nodules were frequently connected by delicate grayish lines.
2. In the heart muscle.
3. Beneath the pleura of diaphragm and lungs.
4. In the mesentery.
5. In or beneath the fascia covering the abdominal wall.

The Malpighian corpuscles in the spleen were frequently enlarged, so that they resembled the pseudo-tubercles very closely.

In addition to the pseudo tubercles there were a few other lesions not usually found in the control animals which indicate a further change in the course of the disease. These variations usually accompany the development of pseudo-tubercles, although they may appear in a less marked form in animals that die presumably before the nodules beneath the serosa are formed. The most important of these variations are as follows:

(a) The purulent infiltration into the subcutaneous tissue at the point of inoculation was less severe than in the control animals.
(b) The areas of necrosis in the liver were less marked, but there was observed a diffuse fatty degeneration of the parenchyma.
(c) The enlargement of the spleen was less marked and the color more normal.
(d) The inguinal and knee-fold lymph glands were more enlarged and the follicles in Peyer's patches more frequently infiltrated.
(e) A membranous exudate was usually noticed over the spleen and liver, frequently accompanied with effusion of serum into the peritoneal and pleural cavities.

B.—RABBITS.

An eruption of tubercle-like bodies was produced on the peritoneum in rabbits by the intraabdominal injection of hog-cholera bacilli attenuated in mixed cultures. The lesions are best described by a case:

February 10, 1891. Two rather small rabbits inoculated with the attenuated hog-cholera culture, one receiving 0.2 cc. subcutaneously, the other 0.15 cc. into abdomen. The former showed no indications of disease at any time; the latter died in nine days.

The autopsy revealed in the abdomen on ventral walls beneath serosa a large number of whitish, slightly elevated, roundish tubercle-like

---

* For cases illustrating the occurrences of these nodules, the reader is referred to the post-mortem notes of guinea-pigs Nos. 3, 10, 13, 17, and 18 on page 47.

bodies 1 to 1.5 mm. in diameter. They are also present in large confluent patches on serosa of cæcum. The Peyer's patches of intestines are infiltrated, the mucosa eroded over some of them. The mesenteric and retro-peritoneal lymph glands contain whitish suppurative foci. In the liver are disseminated numerous minute foci of necrosis. Heart and lungs normal. Some of the tubercles are easily teased out of their surroundings. When broken up, rubbed on cover glasses, and stained numerous mononuclear cells and a considerable number of hog-cholera bacteria appear. In sections of hardened tissue these tubercles appear as a collection of cells under serosa, the center of which shows extensive fragmentation of nuclei. Giant cells are not seen. In the center of many of these foci clumps of hog-cholera bacilli are located. Cultures from this case were plated, and on March 4 a large black rabbit receives into abdomen 0.1 cc. of a bouillon culture twenty-four hours old made from a colony. It was found dead March 13 and showed the following peculiar lesions:

Thorax.—Heart somewhat dilated. Fatty degeneration of walls. Under serosa of lungs very minute translucent tubercles. On tip of right ventral lobe, which is hyperæmic, several large, opaque, whitish tubercles.

Abdomen.—Diaphragm and mesenteries sprinkled with very many minute opacities. On serosa of stomach barely visible tumefactions or tubercles. On spleen a large number of minute grayish tubercles. On liver many barely visible whitish points. Follicles of Peyer's patches infiltrated, mucosa necrosed. The same is true of follicles in appendix of cæcum. Scattering patches of necrosis in cæcum. Suppurative foci in mesenteric glands. Kidneys large with marked parenchymatous degeneration of cortex and cyanotic appearance of medulla. From this case also the injected bacilli were isolated from the spleen.

The foregoing experimental observations lead to certain important deductions, which throw light not only on the subject of preventive inoculation but also on the interrelation of certain groups of infectious diseases.

We have seen that there are certain varieties of the same species of pathogenic bacteria in nature which produce different types of the same disease when inoculated into susceptible animals. We have also seen that the disease produced by the most virulent variety may be so modified by increasing the resistance of the animal as to resemble the various types produced by the more attenuated natural varieties. Lastly, we have demonstrated that some of the types of the inoculation disease as observed in partly immunized animals may be reproduced in fully susceptible animals by the inoculation of an artificially attenuated culture. This relation between the degree of virulence of certain bacteria on the one hand and the relative resistance of the animal body on the other may be expressed by the simple formula—

$$d = \frac{v}{r}$$

in which v = virulence, r = the degree of resistance or immunity of the animal, and d = the type of disease. By changing either virulence or resistance the type is changed. When we reduce the virulence we obtain about the same disease as when we leave the virulence unchanged and increase the resistance of the animal by some kind of preventive treatment, i. e., the value of d remains the same.

Somewhat similar statements have been recently made by Charrin * concerning the pyocyaneus infection in rabbits. He notes much variation in the lesions of any one set of organs, such as the kidneys. His observations, however, pertain more to histological changes, whereas our facts bear on the distribution of macroscopic lesions over the entire body. Charrin injects into an ear vein and thereby at once distributes the disease over the entire vascular system. In many of our experiments the inoculation was merely subcutaneous, and hence the bacteria must be diffused along different routes in order that the peculiar restriction of the disease processes may take place. Thus, in the limitation of the swine-plague bacteria to the pleura after subcutaneous inoculation over the lower part of the abdomen, or in the infiltration and suppuration of Peyer's patches after the subcutaneous injection of hog-cholera bacilli, the route of the bacteria is much more circuitous than in simple intravenous injection, where we may conceive of a complete suppression of the growth of the injected bacteria, excepting at the seat of the lesions. It is not our purpose to attempt an explanation of the phenomena brought out by the immunizing experiments. A complete explanation would presuppose a complete elucidation of the problem of immunity in all its bearings.

A further corollary of much importance may be deduced from these experiments. By increasing the resistance of the smaller experimental animals or by reducing the virulence of the inoculated bacteria we obtain a type of disease simulating that of larger, more resistant species of animals, such as swine. This is very well illustrated in the two diseases experimented upon.

1. By modifying the hog-cholera disease in rabbits so as to prolong it to twice or three times its usual period we obtain instead of the usual septicæmic type, as described on page 10, a true hog cholera as it manifests itself in swine. The lesions become localized in the digestive tract. The solitary and agminated follicles become infiltrated and converted into ulcers and the mucosa of the cæcum may become ulcerated in large patches, as in the cholera of swine. In some rabbits the disease may become localized in the lungs, as is now and then the case in swine.

2. By modifying swine plague in rabbits so as to prolong the disease from twenty hours to a week or more, a localization of the disease in the thoracic organs appears, which corresponds closely with the disease in swine. The bacteria appear to multiply first in the pleural cavity and then to invade the lung tissue secondarily, for the pulmonary lesions may be absent or else restricted to those dependent lobes on which the pleural exudate is most abundant.

These results make it intelligible how much the disease of hog cholera, for example, may vary in pigs in accordance with any variation in the virulence of hog-cholera bacilli and the relative immunity of

---

* Comptes rendus de la Société de Biologie, 1893, pp. 730, 762.

swine, and how difficult it is to decide upon the causation of any particular set of changes in the animal without having recourse to the bacteria involved. We may take it as settled that, in general, swine are not very susceptible to both hog cholera and swine plague. This is proved by the difficulty encountered in producing either disease artificially with pure cultures. The localization of the disease, as compared with the disease in rabbits, also shows this to be true. In other words, when epizoötics of either disease appear among swine they are due largely to aggravating circumstances, and the removal of these is the key to the prevention of swine diseases.

Preventive inoculation, as it has been practiced on domestic animals, does not always insure absolute immunity. Since all immunity is relative and a matter of degree, the resistance of the animal vaccinated will depend upon conditions both internal and external. While vaccination may protect from an acute attack, it may lead to a slower, more chronic disease, not easily recognized, but equally if not more dangerous than the more acute disease, because the animal may become the source of infection for other animals. Such animals are also a loss to the owner, as they are not in a condition to thrive, and will succumb to the chronic affection sooner or later.

Interesting confirmation of these deductions from experimentation with the bacteria of hog cholera and swine plague comes from studies of a European swine disease, *rouget*, or swine erysipelas. This disease exists in nature in about five varieties, and some of these varieties are produced by the process of vaccination. Originally described as a rapidly fatal septicæmia in swine, it has been observed in recent years that it may also appear as an endocarditis. On the inner surface of the heart —usually the left ventricle—wart-like growths are found occupying the mitral valve and almost completely obstructing the flow of blood from auricle to ventricle. In these growths there is a considerable deposit having the character of a thrombus, and in this soft mass rouget bacilli vegetate in large numbers.

This rouget endocarditis has been observed both as a result of vaccination with Pasteur's vaccine and in the course of the natural disease. Hess and Guillebeau * have given some good illustrations of this type of the vaccination disease. Some of the pigs become ill after vaccination and never make a good recovery. In these after death or slaughter the infectious endocarditis is found.

The same form of heart affection was found by Bang † quite common in Denmark. The course of the affection as observed by veterinarians in that country was as follows:

When rouget appeared in a herd of swine it would usually cause more or less acute disease with a certain number of deaths. The

---

* Schweiz. Archiv f. Thierheilkunde, xxviii, 1886.
† Ueber Rothlauf-Endocarditis bei Schweinen. Deutsche Zeitschrift f. Thiermedicin, xviii (1891), p. 27.

remainder of the herd would make a good recovery. One or more months later some would die quite suddenly, the cause of death being infectious endocarditis as a result of the vegetation of rouget bacilli on the valves of the heart. Cases of heart disease were also observed which had not passed through any recognizable acute attack of rouget previously.

Recently C. O. Jensen* described in detail an infectious disease of swine in Denmark, known as urticaria (*Nesselfieber*), which is a modification of the same rouget disease of the continent. At about the same time Dr. Lorenz† described a disease as occurring in Germany which Jensen considers identical with urticaria. It occurs more especially in the warmer season of the year and very rarely ends fatally. At the onset of the disease the pigs become dull, crawl into their bedding, breathe more rapidly, and refuse all food. The temperature rises and may reach (42° C.) 107.6° F. Soon after the onset of these symptoms a peculiar eruption appears on the body, which consists of red, brown, blue, or blackish spots, rectangular in form and sharply outlined (hence the German name *Backsteinblattern*). The longer side of this rectangular reddened spot may be one-half to one inch long, the shorter side about one-third less.

The disease disappears after a few days, but the spots may persist one or two weeks, after which a blackish scab peels off. Lorenz cultivated from one of these red spots a bacillus which closely resembles the bacillus of rouget, both in its biological and pathogenic properties. Rabbits which survived inoculation with rouget resisted successfully inoculation with these bacilli. Jensen's description closely agrees with that of Lorenz. The latter calls attention to the red spots as being not always rectangular, but also roundish or diamond shaped. Jensen in the same article also describes a disease in swine associated with necrosis and sloughing away of large pieces of the skin, sometimes one or more square feet in area. The microscopic examination of such sloughs revealed the presence of large masses of bacilli identical with those of rouget.

Jensen gives the following affections of swine as due to the same bacillus:

1. *Rouget blanc*.
2. Rouget as it is usually known.
3. Diffuse necrotic inflammation of the skin.
4. Urticaria.
5. Endocarditis verrucosa bacillosa.

The careful comparison of cultures from all of these affections, excepting No. 3, of which cultures were not at hand, revealed such slight variations as are common to other species of bacteria.

---

* Die Aetiologie des Nesselfiebers u. der diffusen Hautnekrose des Schweines. Deutsche Zeitschrift f. Thiermed, XVIII (1892), p. 278.

† Archiv. für wiss. u. prakt. Thierheilkunde, XVIII (1892), p. 39.

The form of inoculation disease characterized by the presence of pseudo-tuberculous lesions has certain counterparts in some inoculation diseases among lower animals which have been studied by various observers. In order to trace any possible relationship between the group of hog-cholera bacteria and those that have been described as the cause of experimental pseudo-tuberculosis in rabbits and guinea-pigs the literature of this subject has been examined.

Pseudo-tuberculosis in guinea-pigs was first produced by Malassez and Vingal* and reported in 1883 as "tuberculose zoogloeique." Cultures of the bacteria described as occurring in the tubercles were not made.

Soon after Eberth† described an eruption of submiliary tubercles on the serous membrane of the colon; on the omentum and in the spleen and liver of a rabbit. The tubercles consisted in the main of granulation cells. The bacteria present, according to the description given, resembled hog-cholera bacilli in form and staining. Cultures were not made, however.

In 1888 Charrin and Roger‡ described a pseudo-tuberculous affection produced in rabbits and guinea-pigs by a motile bacillus from 1 to 2 μ long, which multiplies well in ordinary culture media. The bacillus was originally obtained from a guinea-pig which died spontaneously. A subcutaneous inoculation leads to slow emaciation, and death on the thirteenth day. Besides the local swelling, the nearest lymph glands become enlarged, and numerous tubercles are found in spleen and liver. In the lungs they are scarce.

Dor § found what appears to be the same disease independently of Charrin and Roger. The bacillus he isolated contains, in potato cultures, terminal spores. The disease could not be reproduced with cultures, although the pathological products readily produced it. This leaves the impression that the bacillus isolated was not the one causing the tubercles. Grancher and Ledoux,‖ in 1888, obtained from the soil a bacillus which produced lesions similar to those described by former observers as pseudo-tuberculosis. This organism is described as a motile bacillus 1 to 2 μ long, which may, however, assume the form of a very short rod or an ovoid micrococcus. The optimum temperature of the culture is 20° C. This is perhaps the only point in which this bacillus differs appreciably from the colon or hog-cholera group. It should be stated, however, that important cultural characters, among them the action on milk, are omitted. · In the tissues the organism grows in the form of long interlacing chains, of which great masses may appear in certain types of the inoculation disease as the so-called

*Arch. de physiol. normale et pathol., 1883.
†Arch. f. pathol. Anatomie, CIII (1886), p. 488.
‡ Comptes Rendus, 1888–i, p. 868.
§ Loc. cit., p. 1027.
‖ Recherches sur la tuberculose zoogléique. Archives de Méd. Expér., I (1889). p. 209

zoöglœa of Malassez and Vignal. In some types of the disease produced by inoculation the prompt appearance of tubercles may suppress the formation of the zoöglœa. The bacilli may then appear in the tubercles, either isolated or in small masses.

A. Pfeiffer* published in 1889 a short monograph on a form of pseudo-tuberculosis produced by a bacillus originally derived from a glandered horse. Portions of a lung nodule and of a diseased gland were placed under the skin of guinea-pigs with the result that pseudo-tuberculosis and not glanders destroyed the inoculated animals in eight to nine days.

The isolated bacillus differs according to Pfeiffer's detailed careful description only in a few particulars from the hog-cholera group. It is non-motile† and grows preferably in chains. The markings of the surface colony on gelatin plates are also somewhat different from those observed in the hog-cholera and the colon group. Nevertheless the characters as described place this bacillus easily within the hog-cholera group and separate it from the colon group by the absence of any coagulative or other changes in milk.

The lesions observed by Pfeiffer varied somewhat. In the earliest inoculations into guinea-pigs, there was marked infiltration of the inguinal glands with dissemination of nodules in the surrounding connective tissue, in omentum, spleen, and liver. In subsequent inoculations, with death in nine days, the mesenteric glands were greatly enlarged, miliary and submiliary nodules were present in large numbers in spleen and liver. In the spleen there was also a caseous tubercle as large as a pea. The inguinal glands as before.

In still later inoculations, the guinea-pigs died in twenty to twenty-five days. The lesions were still more extensive. To those already described there were added a serous fluid in abdomen, a cheesy deposit on pleura, with a sanguinolent fluid in the pleural cavity, and numerous necrotic foci in the lungs.

By feeding cultures to mice, rabbits, and guinea-pigs, the disease is produced with equal success. The pseudo-tuberculous infiltration is restricted to the lymphatic apparatus of the intestines and to the liver. Among the animals found susceptible are guinea-pigs, wild and tame rabbits, gray and white house mice.

Zagari‡ in 1890, described a form of pseudo-tuberculosis encountered as a spontaneous or natural infection in four guinea-pigs. The isolated bacillus appears in cultures as a non-motile form in long chains, easily broken up, and in shorter oval elements. Its cultural characters so far as given do not differ materially from those of the hog-cholera group. The bacilli appear in masses in the center of the tubercles which con-

---

* Ueber die bacilläre Pseudo-tuberculose bei Nagethieren. Leipzig, 1889.
† See Bacillus η, page 16.
‡ Ueber die sogenannte Tuberculosis "zoogleica" order Pseudo-tuberculose. Fortschritte der Medicin. 1890, pp. 569, 629.

sist of lymphoid elements chiefly. The center of the tubercles may undergo necrosis and caseation. After subcutaneous inoculation of culture fluid a firm infiltration appears at the seat of inoculation which may reach the size of a walnut and subsequently ulcerate. Death ensues in twelve to sixteen days. The neighboring lymph glands are found enlarged. In spleen, liver, and lungs many yellowish, isolated tubercles from the size of a hemp seed to that of a pea are found.

From this brief summary of the bacteriological side of the pseudo-tuberculosis literature we see certain general resemblances between the bacilli found and those belonging to the hog-cholera group. The differences observed are by no means vital. We do not wish to argue for or against any close relationship, however, until more data are at hand. The production of tubercles with attenuated hog-cholera bacteria or with virulent bacteria in partly immune animals is of sufficient importance to warrant the above summary. Further study of spontaneous pseudo-tuberculosis might perhaps lead to a better understanding of the origin of the hog-cholera group of bacteria.

The foregoing investigations throw much light upon the possible variations of all infectious diseases, both human and animal. We need but recall here the great variety of lesions produced by the ordinary septic and pyogenic bacteria of man, the staphylococci and the strep-tococci. Furthermore, most specific infectious diseases show variations in type, variations in the localization of the most distinctive pathological alterations. Some obscure forms of disease still needing elucidation may perhaps be recognized in the future as modifications of such as are already understood. As illustrations may be cited the recent discussions as to whether certain obscure diseases of man are of leprous origin or not, or whether scarlatina is a simple form of streptococcus infection. The outcome of our work emphasizes the importance of renewed continuous etiological studies. In infectious animal diseases this is of the utmost importance, for the unrecognized existence of some form of an infectious disease favors its unchecked dissemination. The confusion which has prevailed on the subject of infectious swine diseases must be attributed largely to the many possible variations in the symptoms and lesions as determined by post-mortem examinations. Such variations are due on the one hand to the condition of the animal as regards individual resistance, age, breed, the state of the internal organs, the amount of fat, etc., on the other to the virulence of the specific bacteria. The only final test of the nature of the disease is the character of the bacteria responsible for it.

CAN THE BACILLUS OF HOG CHOLERA BE INCREASED IN VIRULENCE BY PASSING IT THROUGH A SERIES OF RABBITS?

By Veranus A. Moore.

In 1890 Selander [*] reported the results of certain experiments with the bacillus of *Svinpest* (Danish hog cholera), in which he showed that its virulence could be rapidly accelerated by passing it through a series of rabbits or pigeons. He furthermore affirmed that the germ was identical with the bacillus of American hog cholera. This most remarkable result, which was quite contrary to the conclusions which had been reached from the work in this laboratory, gave rise to the presumption that his culture had inadvertently become contaminated with the swine-plague or some other organism more virulent than the bacillus with which he was working, and consequently the question was temporarily dismissed.

In 1892 the results of Selander were confirmed by Metchnikoff [†] in an article on certain experiments concerning the immunizing properties of the products of hog-cholera bacteria. As no extended experiments had been made in this laboratory in precisely the line followed by Selander, it became necessary after Metchnikoff's publication to determine more positively whether or not the bacillus of hog cholera was susceptible to such rapid changes in its virulence. That hog-cholera bacteria of different degrees of virulence exist in nature had already been demonstrated by inoculation experiments with the bacilli isolated from different outbreaks of that disease. The tenacity with which these bacteria retained the virulence which they possessed at the time of their isolation, when cultivated under ordinary conditions or by passing them through experimental animals, had also been observed. Many efforts had been made to change the virulence of these bacteria, but invariably the results showed that it could not be quickly done by any of the methods employed. [‡] The double assurance, therefore,

---

[*] Annales de l'Institut Pasteur, IV (1890,) p. 543.

[†] *Ibid.*, VI (1892), p. 289.

[‡] *See* Reports of the Bureau of Animal Industry since 1885.

from European investigators that the virulence of the American hog-cholera bacillus could be rapidly increased by passing it through a series of rabbits was somewhat startling. Either to verify their results which appeared at the time to be of much significance, or to acquire additional evidence to support the theory that the virulence of this bacillus is not subject to such artificial changes, an experiment on rabbits was made.

After this experiment was completed it was found that the germ with which Metchnikoff had worked was the bacillus of swine plague * and not of hog cholera, as stated in his article. As he confirmed Selander's experiments, it was not unreasonable to suppose that Selander had made the same error. We have, however, no positive evidence of this, as a culture of his germ has not been studied, and the presumption of an error in the identification of the organism is necessarily based solely on the results he reported and on the statements of Metchnikoff. As the bacillus of the disease in Denmark was described by Selander,† and according to his description it closely resembles the bacillus of hog cholera in form, motility, growth in gelatin, and appearance in animal tissues, it seems unjust to affirm, on the evidence cited, that he had made a mistake in the identification of a germ which he himself had described. On this account his experiment is given that his results may be compared with those obtained in a similar experiment with the bacillus of American hog cholera.

His experiment was made for the purpose of increasing the virulence of the bacteria of *Svinpest* (hog cholera) by passing them through a series of rabbits and pigeons. The germ which he used had been preserved by means of subcultures in gelatin for about two years. It was originally obtained by Selander from an organ of a pig. After this long cultivation it required 1 cc. of a bouillon culture to destroy a rabbit in three days; it would not kill pigeons. With this virus Selander conducted the following experiments:

The spleen of a rabbit which perished in three days from a subcutaneous inoculation with a pure culture was made into an emulsion and a second rabbit inoculated subcutaneously with 1.5 cc. of this emulsion;‡ a third rabbit was inoculated with a similar quantity of an emulsion made with the spleen of the second rabbit, and a fourth from that of the third. The fourth rabbit died in fourteen hours. Three pigeons were inoculated subcutaneously with 0.1, 0.25 and 0.5 cc., respectively, of the emulsion of the spleen of the fourth rabbit. They perished in fourteen, five. and seven days, respectively. With the blood of the first pigeon, which had been in the incubator for several hours in order that the bacteria could multiply, a fourth pigeon was inoculated (five drops of the blood mixed in a little bouillon was injected); it

---

* At Dr. Smith's request Metchnikoff very kindly sent him a culture of the bacillus with which he had worked. It was found to be identical with the swine-plague germ. For a description of this organism see footnote, p. 60, in this publication.

†Centralblatt für Bakteriologie und Parasitenkunde. Bd. III (1888), p. 362.

‡He does not state how much or what liquid was used to make the emulsion, and consequently the strength of the dose can not be estimated.

perished in thirty-six hours. He continued the passage from pigeon to pigeon until 148 pigeons had been inoculated. The fifth pigeon perished in less than twelve hours. In subsequent cases death occurred in from eighteen to thirty-six hours. Later the irregularities became less and less, and the virulence of the germ fixed, so that a pigeon inoculated at 6 in the evening with from 0.05 to 0.2 cc. of the defibrinated blood of the preceding one would be found dead on the following morning. A very few of the pigeons resisted for a few days. In these, few bacteria were found in the blood, but each exhibited pericarditis with a fibrinous exudate rich in bacteria. The bacteria which were exalted in their virulence for pigeons were also rapidly fatal to rabbits. A subcutaneous inoculation with 0.01 to 0.25 cc. destroyed rabbits in from twelve to fifteen hours, and a rabbit inoculated in the ear vein with 0.05 cc. of virulent blood perished in five hours. The blood from the pigeons was also rapidly fatal to swine. A pig nine weeks old received 0.5 cc. of the blood from pigeon No. 37 in the ear vein at 2 o'clock on the morning of April 3. It was found dead the following day.

It was my purpose to repeat Selander's experiment in every detail in order that a more accurate comparison of the results could be made. A few omissions in the details of his work, however, rendered it impossible to determine certain minor points in his process. The history of the bacillus which I used and the method that was followed in this experiment are appended.

The bacillus was obtained in pure culture from the spleen of a pig which died of hog cholera in La Salle County, Ill., in the fall of 1891. It had been preserved by means of subcultures on agar for about eight months when this experiment was begun. It had not lost its virulence to any appreciable degree. Rabbits inoculated subcutaneously with 0.1 cc. of a fresh bouillon culture would die in from five to eight days. It should be stated that hog-cholera bacteria have rarely been found in the investigations of this laboratory that were sufficiently virulent to destroy rabbits in less than three days when they were inoculated subcutaneously with not more than 0.1 cc. of a fresh bouillon culture. In the preparation of the emulsion from the spleen of the rabbits the following order was observed:

The rabbits usually died during the night. They were examined sometime between 9 and 11 o'clock on the following morning. The spleen was removed very carefully to prevent contamination from without. It was cut into small pieces with sterile scissors and then ground up in a sterilized mortar with 10 cc. of sterile bouillon. Of this emulsion 0.5 cc. was used for each inoculation. Gelatin roll cultures were made from the emulsion in several instances. These developed as many colonies as would usually appear in similar rolls made from fresh (twenty-four-hour) bouillon cultures. The appended table contains all necessary information concerning these inoculations:

Series of inoculations in rabbits with hog-cholera bacteria.

Rabbit number—	Weight of rabbit.	Date of inoculation.	Rabbit inoculated subcutaneously with—	Date of death.	Time rabbit lived after inoculation.	Remarks.
	Grams.				*Days.*	
1	1,482	May 23, 1892	0.25 cc. bouillon culture, hog-cholera bacteria.	May 28	5	Usual hog-cholera lesions.
2	1,482	May 28, 1892	0.5 cc. emulsion spleen, rabbit 1...	June 2	4	Ecchymoses beneath pleura and pericardium.
3	1,368	June 2, 1892	0.5 cc. emulsion spleen, rabbit 2...	June 7	5	Usual hog-cholera lesions.
4	1,216	June 7, 1892	0.5 cc. emulsion spleen, rabbit 3...	June 11	4	Lower colon and duodenum hemorrhagic.
5	1,596	June 11, 1892	0.5 cc. emulsion spleen, rabbit 4...	June 18	8	Usual hog-cholera lesions.
6	1,707	June 18, 1892	0.5 cc. emulsion spleen, rabbit 5...	June 21	3	Very many coccidia cysts in liver.
7	1,140	June 21, 1892	0.5 cc. emulsion spleen, rabbit 6...	June 29	8	Very severe local reaction; puncitiform hemorrhages in peritoneum.
8	1,596	June 29, 1892	0.5 cc. emulsion spleen, rabbit 7...	July 5	6	Usual hog-cholera lesions.
9	1,672	July 5, 1892	0.5 cc. emulsion spleen, rabbit 8...	July 11	6	Do.
10	1,140	July 11, 1892	0.5 cc. emulsion spleen, rabbit 9...	July 15	4	Do.
11	1,425	July 15, 1892	0.5 cc. emulsion spleen, rabbit 10..	July 22	7	Do.
12	1,596	July 22, 1892	0.5 cc. emulsion spleen, rabbit 11..	July 27	5	Do.
13	2,280	July 27, 1892	0.5 cc. emulsion spleen, rabbit 12..	Aug. 1	5	Do.
14	1,710	Aug. 1, 1892	0.5 cc. emulsion spleen, rabbit 13..	Aug. 6	5	Do.
15	1,425	Aug. 6, 1892	0.5 cc. emulsion spleen, rabbit 14..	Aug. 12	6	Do.
16	1,425	Aug. 12, 1892	0.5 cc. emulsion spleen, rabbit 15..	Aug. 17	5	Do.
17	2,223	Aug. 17, 1892	0.5 cc. emulsion spleen, rabbit 16..	Aug. 24	7	Do.
18	1,824	Aug. 24, 1892	0.5 cc. emulsion spleen, rabbit 17..	Aug. 30	6	Spleen unusually small.
19	1,026	Aug. 30, 1892	0.5 cc. emulsion spleen, rabbit 18..	Sept. 3	4	Spleen apparently normal.
20	1,710	Sept. 3, 1892	0.5 cc. emulsion spleen, rabbit 19..	Sept. 9	6	Usual hog-cholera lesions.
21	1,824	Sept. 9, 1892	0.5 cc. emulsion spleen, rabbit 20 .	Sept. 14	5	Do.
22	2,280	Sept. 14, 1892	0.5 cc. emulsion spleen, rabbit 21..	Sept. 19	5	Do.
23	1,710	Sept. 19, 1892	0.5 cc. emulsion spleen, rabbit 22..	Sept. 26	7	Do.
24	1,254	Sept. 26, 1892	0.5 cc. emulsion spleen, rabbit 23..	Oct. 1	5	Do.
25	1,311	Oct. 1, 1892	0.5 cc. emulsion spleen, rabbit 24..	Oct. 6	5	Do.
26	1,596	Oct. 6, 1892	0.5 cc. emulsion spleen, rabbit 25..	Oct. 11	5	Do.

The table shows that a series of inoculations with the hog-cholera bacillus, including 26 rabbits and extending over a period of more than four months time, had no effect whatever in increasing its virulence. In fact, the average time required for the inoculation to kill a rabbit in the first half of the experiment was a trifle shorter than that required with the last ten animals, which indicates a tendency toward attenuation rather than exaltation of the virus. The more rapid death of rabbit No. 6 is explained by the severe invasion of coccidia in the liver. The slight difference in the time required for the inoculation to produce death in the different rabbits can be accounted for on the ground of individual susceptibility or resistance.

Selander's germ was said not to be virulent enough to destroy pigeons, but that the emulsion of the spleen of the fourth rabbit would do so when inoculated in very small quantities. The bacillus of hog cholera which I used in making the rabbit experiment would occasionally kill pigeons when 0.5 cc. of a bouillon culture was injected into the pectoral muscle. This would indicate that his germ was less destruc-

tive to pigeons than the one I employed. However, a series of inoculations in pigeons was attempted in order to complete the comparison
and to test the effect on the virulence of the hog-cholera bacillus by
passing it through a series of pigeons. That a slight degree of increase
in the virulence of the bacillus produced in this way might be detected,
a few pigeons were inoculated with a pure culture of the bacillus made
directly from the culture from which the first rabbit in the series was
inoculated. Pigeons were then inoculated with certain quantities of
the emulsion from the spleen of different rabbits in the series, to determine whether the bacillus had become more virulent for pigeons than
it was at the beginning of the rabbit experiment. The result of the
preliminary inoculations into pigeons is given in the following table:

Inoculation of pigeons with hog-cholera bacteria.

Pigeon number—	Date of inoculation.	Method of inoculation.	Virus used.	Results.
1	May 25, 1892	In pectoral muscle..	0.1 cc. bouillon culture..	Sick; recovered.
2	...do	....do	0.25 cc. bouillon culture.	Died in 22 hours.
3	...do	....do	0.5 cc. bouillon culture..	Died in 36 hours.
4	May 27, 1892	Beneath the skin....	0.25 cc. bouillon culture.	Sick; recovered.
5	...do	....do	0.5 cc. bouillon culture..	Do.
6	June 11, 1892	....do	0.5 cc. emulsion spleen, rabbit No. 4.	Do.
7	...do	....do	0.25 cc. emulsion spleen, rabbit No. 4.	Remained alive; unthrifty.
8	...do	....do	0.1 cc. emulsion spleen, rabbit No. 4.	Do.
9	...do	In pectoral muscle..	0.25 cc. emulsion spleen, rabbit No. 4.	Do.
12	June 20, 1892	...do	0.5 cc. bouillon culture ..	Died in 24 hours.
13	June 23, 1892	Beneath the skin....	0.5 cc. blood pigeon 12...	Remained well.
14	...do	....do	0.75 cc. blood pigeon 12..	Died in 4 days.

The uncertainty with which even a large quantity of the culture or
emulsion of the spleen of the rabbits, or of the pectoral muscle of the
pigeon would kill pigeons, rendered a consecutive series of inoculations impossible. After several efforts to obtain such a series, in which
a large number of pigeons were used, the experiment was abandoned.
Certain of the pigeons that died from the inoculation were placed in an
incubator for from six to twenty-four hours that the hog-cholera bacteria might multiply before the emulsions were made from the pectoral
muscle or heart blood, but no additional virulence was obtained. A
series of inoculations into the pectoral muscle which was begun with
the emulsion of the spleen from rabbit No. 15 gave promise of success,
but after a few successful inoculations the virus became contaminated
with *Bacillus fluorescens liquefaciens.* * The mixed culture was more
virulent, but after isolating the bacteria by means of agar plates the
hog-cholera bacillus failed to kill pigeons when 0.5 cc. of a bouillon

---

* The *B. fluorescens liquefaciens* was fatal to pigeons when 0.5 cc. of a fresh bouillon
culture was injected into the pectoral muscle. This is the second time that I have
found this species of bacteria possessed of pathogenic properties. It is very frequently isolated from diseased animal tissues, the secretions covering the mucous
membranes and from various extraneous material.

culture was injected into the pectoral muscle, showing that its virulence had not been appreciably increased.

The results obtained in these series of inoculations are important in
demonstrating the constancy of the virulence of the hog-cholera bacillus when treated in this manner. They fully verify the results of more
fragmentary experiments heretofore made with this bacillus and show
conclusively that the results obtained by Selander do not hold true
for American hog-cholera bacteria.

In the preceding article it was shown that by substituting the words
swine plague for hog cholera in Metchnikoff's work his results could be
verified. In the same article (p. 67) it was found that the swine-plague
bacteria obtained from the abscess in the abdominal cavity of rabbit
No. 16 were attenuated, but that they were restored to their original
virulence by passing them through a series of three rabbits.* In
this case, however, the rabbits were inoculated with fresh cultures
instead of emulsions from the spleen. This single experiment with
virulent swine-plague bacteria, which had become temporarily attenuated by their long life in the abscess, is in harmony with Selander's
results. The conditions are also comparable, as his bacillus was at one
time virulent, but had become attenuated through long cultivation. It
is a significant fact that while he carried on a long series of inoculations his bacillus reached a high degree of virulence in three passages,
as the fourth rabbit died in fourteen hours. These results indicate
that if we substitute swine plague for hog cholera and *Svinpest* in
Selander's article his work also can be verified.

The final and most important conclusion to be drawn from the foregoing experiment is the reply to the question with which it was begun.
To this there appears to be but one answer, namely, that the bacillus
of hog cholera as encountered in swine can not be rapidly or even
slowly increased in virulence by passing it through a series of rabbits.

---

* In the summer of 1893 a similar series of inoculations were made with the
bacillus of rabbit septicæmia, which was obtained from an outbreak of that disease.
No appreciable increase in the virulence of the germ was obtained. This bacillus
was very attenuated, producing extensive pleuritis and death in from 2 to 4 days
after an intravenous inoculation of 0.4 cc. of a fresh bouillon culture. This indicates that even germs belonging to the swine-plague group of bacteria which
appear in nature in an attenuated condition are not so susceptible to artificial
changes in their virulence as those which appear in a virulent form and are subsequently attenuated.

WHAT BECOMES OF HOG-CHOLERA AND SWINE-PLAGUE BACTERIA INJECTED IN SMALL NUMBERS INTO THE SUBCUTANEOUS TISSUE OF PIGS?

By Veranus A. Moore.

In the investigation of swine diseases it became desirable to determine the extent to which hog-cholera and swine-plague bacteria would be disseminated through the body of pigs when a small quantity of a pure culture of these organisms was injected beneath the skin. It was likewise important to determine the time during which these bacteria would remain alive in the organs of a healthy animal which had been inoculated.

EXPERIMENT WITH HOG-CHOLERA BACTERIA.

In the experiment with the hog-cholera bacillus four well-nourished pigs (Berkshire crosses) two and one-half months old and weighing from 45 to 60 pounds were selected. Two of them received subcutaneously, by means of a hypodermic syringe, on the inside of the right thigh, 1 cc. each, and the other two, 1.5 cc. each, of a bouillon culture of hog-cholera bacteria. The culture used was prepared by inoculating a tube of peptonized bouillon with hog-cholera bacteria and placing it in an incubator for two days, when it was removed and allowed to stand in the laboratory at the room temperature for three days more before it was used. The growth was vigorous and the bacteria still motile. A culture that had grown for five days was employed because in former experiments it had been noticed that a culture increases in virulence up to a certain age.

The pigs were killed at different intervals, and a very careful examination was made of all the organs, excepting the brain. In order to determine approximately the number of bacteria present in the different tissues a gelatin roll culture was made from each. As a check upon the roll culture, a tube of peptonized bouillon was inoculated from the same organs. For the sake of comparison, care was taken to use as nearly as possible the same quantity of tissue for each culture; of the solid organs a piece about the size of a small bean was taken, and of the blood two large loops. The tissues were thoroughly crushed and

103

broken by means of sterilized forceps. The tissues from which cultures were made from each animal were: the place of inoculation, blood from the heart, lungs, spleen, liver, kidney, lymphatic gland at the smaller curvature of the stomach, and the bronchial gland. Stained cover-glass preparations from the blood and spleen showed no bacteria; those from the local lesions in the first three animals that were killed contained a large number of hog-cholera bacteria, but in those from the fourth no germs could be found. The result of these examinations, together with the history of the inoculations, are summarized in the subjoined table:

Table showing the distribution of hog-cholera bacteria in the organs.

Pig number—	Date of inoculation.	Culture injected subcutaneously.	Pig killed—	Days after the inoculation.	Hog-cholera bacteria in bouillon cultures from—	Number of colonies of hog-cholera bacteria in the gelatin rolls.	Other bacteria in bouillon cultures from—
77	Nov. 3, 1891	cc. 1	Nov. 5, 1891	2	Local lesion.....	1,000–2,000	Bronchial gland, blood and liver. a
80	...do	1.5	Nov. 10, 1891	7	Local lesion..... Bronchial gland. Stomach gland..	1,000–2,000 20 50	Liver. a
78	...do	1.5	Nov. 14, 1891	11	Local lesion.....	1,000–2,000	Bronchial gland.
79	...do	1	Dec. 5, 1891	30	Bronchial gland. Stomach gland.. No hog-cholera bacteria.	No colonies. No colonies. No colonies....	Bronchial gland.

a A pure culture of an anaërobic, spore-bearing bacillus.

It is of interest to note that the bacteria were not found beyond the place of injection two days after the inoculation, and that in eleven days they did not develop in the gelatin rolls, excepting in the one made from the local lesion. It will be observed, however, that the hog-cholera bacteria were disseminated through the body, for they were detected in certain of the lymphatic glands five and eleven days after they were injected into the subcutaneous tissue. In thirty days they had disappeared from these glands and from the place of inoculation.*

---

* In the summer of 1893 a somewhat similar experiment was made by Dr. Smith for another purpose, but the results obtained are of interest in this connection. June 5, pig No. 144 was inoculated subcutaneously in the left iliac region with 1 cc. of a bouillon culture of hog-cholera bacteria. A swelling about 6 cm. in diameter developed at the point of inoculation. In the center of this there was a firm indurated nodule about 2 cm. in diameter. The swelling subsided in a few days, but the induration remained as a firm nodule for several weeks. July 8, it was reinoculated on the right side with 1 cc. of a similar culture. The local induration which followed was more severe than after the first inoculation. August 10, the pig was again reinoculated with a similar quantity of the hog-cholera culture. This time a small nodule developed at the point of inoculation. The pig was killed September 12, thirty-three days after the last inoculation, and tubes of bouillon and agar were inoculated with pieces of the local induration caused by the last inoculation, and with pieces of the liver. These remained clear, and cover-glass preparations made from the local induration showed no bacteria.

It is an interesting fact that in no case hog-cholera bacteria were found in the blood, spleen, liver, or kidneys, where they are almost invariably detected when the animals perish from hog cholera contracted either by exposure to the disease or by inoculation. It is unfortunate that a few additional cases could not have been added between the third and the fourth animal so that the time of the disappearance of the germs could have been more exactly determined.*

These inoculations demonstrate the fact that when hog-cholera bacteria are injected in small numbers they are taken to different parts of the body and harbored for a considerable time in certain lymphatic glands as well as in the subcutaneous tissue at the place where they were injected, and that eventually they are destroyed. There is very little, if any evidence, that the bacteria multiplied to any great extent within the animal body. The swelling at the place of inoculation can not be attributed solely to this cause, as injections of sterilized cultures have frequently been found to produce severe local reaction.

The other bacteria obtained from the bronchial glands were ordinary aërobic saprophytic organisms; the same is true of the germ found in the blood of pig No. 77. The anaërobic spore-bearing bacillus obtained from the liver of pigs Nos. 77 and 80 probably belonged to the group of bacteria containing *Bacillus butyricus*. It grew only in the bottom of bouillon tubes and died out after the second subculture was made.

The animals were killed by a blow on the head. This may account for the presence of bacteria in the blood of pig No. 77, as this animal lived some minutes after the blow was administered, giving the blood a chance to carry bacteria from the injured parts. The examination showed no lesions due to the injection of the virus, excepting a local swelling,† and at no time did the pigs show any ill effect from the injection of the bacteria. The local reaction as shown by the post mortem was somewhat variable, owing to the difference in the time of death after the inoculation.

A rabbit was inoculated subcutaneously with 0.1 cc. of the bouillon

---

*The considerable number of colonies of hog-cholera bacteria that developed in the roll cultures made from the lymphatic glands after five days, and their absence in the gelatin rolls made from the glands after eleven days had elapsed, indicate that the destruction of the bacteria in the lymphatics was nearly completed in that time.

† In pig No. 77 there was a purulent infiltration and thickening of the subcutis extending over an area 5 by 3 cm. In pig No. 80 the subcutis was thickened, quite firm and apparently necrosed over an area 8 by 2 cm. In pig No. 78 there was a quite firm tumor about 6 by 2 cm. underneath the skin. The exterior layer of the tumor was of a pale, pinkish, translucent appearance, while the interior was firm and whitish, due to a dense infiltration of leucocytes. It was attached to the subjacent muscle by loose connective tissue. In pig No. 79 there was a scar at the point of injection beneath which was a layer of purulent substance about 2 mm. thick, and extending over an area about 5 cm. in diameter. Around this the connective tissue was slightly thickened.

culture from the bronchial gland of pig No. 80. It died of hog cholera on the fourth day.

EXPERIMENT WITH SWINE-PLAGUE BACTERIA.

The experiment with swine-plague bacteria was carried out in a manner similar to the one with hog cholera just described. A few differences in the details, however, should be mentioned. Two pigs were selected from the same herd as those taken for the previous experiment. They were inoculated subcutaneously in the right thigh with 1 cc. each, of a turbid suspension in bouillon of the surface growth from an agar culture of swine-plague bacteria one day old.* As the swine-plague bacillus grew very unsatisfactorily in gelatin the cultures from the various organs were made on the inclined surface of agar and in bouillon. In other respects the details were the same as those already described in the experiment with hog-cholera bacteria. A summary of the inoculations and the result of the bacteriological examination of the various organs is given in the subjoined table:

Table showing the distribution of the swine-plague bacteria in the tissues.

Pig number—	Date of inoculation.	Bacteria injected subcutaneously.	Pig killed—	Days after inoculation.	Swine-plague bacteria in cultures from.	Colonies of swine plague on surface of agar.	Other bacteria in cultures from.
85	Nov. 14, 1891	1 cc. suspension agar cultures.	Nov. 16	2	Local lesion only.	So numerous as to be confluent.	Kidney. Bronchial glands and lung.
81	....do	....do	Nov. 20	6	....do	....do...........	Bronchial gland and liver.[1]

[1] Anaërobic spore-bearing bacillus.

From this experiment it is evident that the swine-plague bacteria were either not disseminated through the body, as in case of the hog-cholera bacilli, or that being taken up they were destroyed more rapidly than were the hog-cholera bacteria.

There were no disturbances produced by the inoculations, and the post-mortem examinations revealed no lesions that were produced by the injections, excepting the local reaction. This varied somewhat; in pig No. 85 there was a thickening of the subcutaneous tissue about the point of injection extending over an area 4 cm. in diameter; the subjacent muscle was roughened; considerable blood extravasation in the thigh muscle. In pig No. 81 there was a small (1.5 cm. in diameter) tumor beneath the skin at the place of inoculation consisting of a soft, grayish nucleus surrounded by a quite firm connective tissue wall.

From the cultures and cover-glass preparations it was impossible to determine whether there was any difference in the number of bacteria

---

*It should be noted that the turbid suspension from the agar cultures of the swine-plague bacteria injected was equivalent to a much larger quantity of a simple bouillon culture so far as the number of bacteria is concerned.

in the local lesion of the two animals produced by the injection of the bacteria. The tubes of agar inoculated developed confluent growths, and in stained cover-glass preparations no bacteria could be detected in either case. It is presumable that there was at most only a slight multiplication of the bacteria in the subcutaneous tissue. This is indicated by the somewhat diffuse lesion in the pig killed in two days and the formation of a circumscribed tumor at the seat of inoculation in the animal that was allowed to live six days.

The identity of the bacillus obtained from the local lesion of pig No. 81 was established by inoculating a rabbit with a very small quantity of the culture. It was found dead on the following morning and its organs contained innumerable swine-plague bacteria. The spore-bearing bacillus found in the bouillon culture from the liver of pig No. 81 was apparently identical with those found in cultures from the livers of pigs Nos. 77 and 80.

It is instructive to note that in 5 of the 6 pigs used in these experiments the bronchial gland was found to be infected with various species of bacteria. This fact is valuable as it demonstrates that bacteria may be carried through the lungs into the corresponding lymphatic glands.

From these experiments the following conclusions appear to be warranted concerning the fate of these bacteria when injected subcutaneously in small numbers.

1. Both hog-cholera and swine-plague bacteria will remain alive in the subcutaneous tissue for several days after their injection.

2. The hog-cholera bacteria are taken up from the point of injection and distributed, to a certain extent, through the body. They are harbored for a limited period of time in certain lymphatic glands where they may be detected. They are not found in the other organs of the body.

3. The swine-plague bacteria are not found beyond the tissues immediately surrounding the point of their injection.

4. Subcutaneous injection of small doses of either hog-cholera or swine-plague bacteria of ordinary virulence have little, if any, pathogenic effect.

By THEOBALD SMITH.

The experience of the past fifteen years has amply shown that all facts concerning the nature of important disease germs, however insignificant they may appear by themselves, should be worked out and recorded. Slight variations or modifications of such potent organisms may express themselves in the shape of extensive epidemics and epizoötics, for it seems to require but a slight elevation in the virulence of any pathogenic organism to greatly extend its injurious operations. The careful study of the hog-cholera group of bacteria is thus a matter of necessity. Thus far we have found a well-defined bacillus associated with outbreaks of hog cholera in a considerable number of localities, both in the East and the West. The relation of this bacillus to the disease is unquestioned. Furthermore, this bacillus occurs under a number of varieties, whose chief distinguishing character lies in the varying degree of pathogenic activity. Some are more, others less virulent, and the type of disease produced in swine is correspondingly varied. The outbreak from which a few cases are reported in full on page 27, and from which a well-defined variety was isolated, differs in important particulars from other types of uncomplicated hog cholera in the greater duration of the disease and the peculiar diphtheritic character of the inflammation of the intestines. The confusion existing even to-day in the minds of many authorities on the subject of swine diseases, expressing itself in the multitudinous names which have been given to them, evidently rests in part on the different types which hog cholera assumes as a consequence of the pathogenic variability of the bacillus.*

Besides those varieties of this bacillus actually encountered in swine diseases there are other more or less closely related varieties of this species or group of species which have been found in connection with disease among other domestic animals.

In Bulletin No. 3 I described a bacillus from the vagina of a mare after abortion, which certainly is a hog-cholera bacillus. In Europe, Gærtner found a bacillus in the organs of a sick cow, the meat of which

---

* In the bulletin on Swine plague (1891) a study of the varieties of the swine-plague bacillus will be found.

caused a widespread epidemic of gastro-intestinal disease in man. In the foregoing pages I have recorded facts which show that this bacillus is hardly more than a variety of the hog-cholera bacillus. The same may be said of the mouse-typhus bacillus of Loeffler, which the discoverer has recommended and used in the destruction of field mice and other pests.* In the observations given in the preceding pages this organism is shown to possess pathogenic properties of a high degree, but differing somewhat from those possessed by the true hog-cholera bacillus.

Both these organisms were found in Germany, yet there is no positive evidence that hog cholera exists there. It is true that neither organism is completely identical with the virulent hog-cholera bacillus as found in this country, yet they approach quite closely the less pathogenic varieties. This leads us to the broad question whether the different methods of rearing and feeding swine have led in the one country to a repression of such diseases, and in the other to an extension of them by reducing the resistance of the digestive organs of swine. The cultivation of the hog-cholera bacillus by the continuous existence of the disease may have finally led with us to races of maximum virulence.

Intimately associated with the problem of variations in this group is the one dealing with the possible extension of hog cholera to other domestic animals. This must be constantly kept in view, and attention has already been called to this phase of the hog-cholera problem in Bulletin No. 3. Thus far members of this group have been found in a cow, in mice, and in a mare after abortion. Even if the hog-cholera bacillus may not at present produce epizoötics in other animals, the possibility is open that it may produce various disturbances in isolated cases, that it may act with other disease germs, and, finally, that it may slowly adapt itself to other species.

This adaptation must be a slow process, for the foregoing investigations show how tenacious the characters of the hog-cholera bacillus are and how difficult it is to modify its virulence, either by methods of cultivation or passage through animals. This fact, which has been brought out in various ways in former investigations, has been confirmed by the inoculation experiments reported by Dr. Moore on page 100. This species is, on the whole, very stable, and its very stability is responsible for its dangerous properties by putting obstacles in the way of preventive inoculation, disinfection, and the recovery of diseased animals.

In carrying on vaccination experiments on the smaller experimental animals we were guided by the fact that the problem of immunity is to-day the foremost one in the more abstruse study of infectious diseases, and that its ultimate solution will furnish efficient practical ideas concerning those causes which predispose animals to infection. The study of any one infectious disease will not lead us very far. A num-

* Annual Report of the Secretary of Agriculture for 1893.

ber of different maladies must be thoroughly exploited. For us the immediate task appears to be the investigation of the problem in connection with those animal diseases which are of the greatest economic importance and about which we had gathered together much information. In other countries the same problems have been attacked in connection with other diseases, and by mutual coöperation much advantage has accrued. In the experiments reported in these pages some important facts have been disclosed. We have learned that swine, as a rule, are not very susceptible to either disease. This is proven by the difficulty of producing disease either by feeding cultures or by injecting them under the skin. It is also shown by the nature of the diseases as they manifest themselves in swine. They are always localized in one or more organs, while in the very susceptible small animals both diseases appear as generalized septicæmic diseases of a speedily fatal character and are produced by inoculation with exceedingly minute doses. If swine therefore already possess a fair degree of immunity against these diseases we must look for accessory causes to account for the extensive mortality. Those conditions which favor an unhealthy condition such as monotonous, badly chosen food rations and animal parasites, especially lung worms and ascarides, have already been discussed in the bulletin on swine plague, as possible agents in breaking down those protective barriers to disease which in the swine diseases consist mainly in perfect digestion and healthy lungs. Swine should receive at least some of the attention and care that is given to other domestic animals, and sanitary laws can not be set aside without inviting disease germs to enter the system. It is taken for granted that proper precautions are to be taken at all times to prevent the introduction of sick or exposed animals into a herd, but there may be regions where the surroundings, the soil, and the water courses are being so continually infected that the disease germs may be presumed to be present at all times, and where the only safeguards against infection will be a proper amount of attention to the general health of the animals and to the removal of those conditions which are likely to impair it.

How important a part the food plays in predisposing animals to hog cholera and swine plague has not thus far been studied. Certain it is that such one-sided food as corn and water can not be recommended. It has been shown that when rats are fed on a flesh diet they are more likely to resist anthrax after inoculation than when fed on bread.* When those fed on bread also received salt, increased resistance was likewise observed. A reduction in the necessary quantity of proteid food with swine may perhaps be responsible for more injury than appears on the surface. Errors in feeding are not likely to limit their injurious activity to one disease alone but may increase susceptibility to disease in various directions. Improper food besides reducing the general resistance of the body cells to bacterial invasion may lead to

*Kurt Müller: Der Milzbrand der Ratten. Fortschritte der Medicin, 1893, p. 522.

a weakening of certain important organs such as the stomach. It has been frequently shown how important normal digestion is for the prompt destruction of bacteria in the food. A catarrhal condition inducing a less acid reaction of the mucous membrane may permit hog-cholera bacteria to pass through this organ alive and once in the small intestine they will invariably set up disease.

Our experiments have furthermore shown that the different degrees of susceptibility may produce quite a variation in the character of the disease, and that when a certain degree of immunity has been produced in rabbits it would be impossible to foretell what form the inoculation disease would assume. It might become localized in the intestines or the lungs, or even in the brain. Swine plague might appear as a peritonitis or a pleuritis with or without pneumonia, or in the form of one or more abscesses. This interesting fact is at present of most service to the investigator, because he will be better able to identify obscure diseases. It is needless to insist on the importance of this knowledge. The presence of mild, chronic hog cholera or swine plague in a herd may be fully as dangerous as the more acute disease, for mild disease is frequently caused by highly virulent bacteria. This has been fully brought out in the foregoing pages. Rabbits which, through vaccination, have lived many months after inoculation have continued to carry disease germs in their body virulent enough to kill unvaccinated rabbits in twenty hours.

Another fact of considerable importance is the relative character of the immunity produced by vaccination. An animal which before preventive inoculation possessed a high degree of susceptibility may after treatment no longer contract an acute rapidly fatal disease but one of a more chronic character, which dragging itself along for months and rendering the animal worthless may form the starting point of subsequent outbreaks among newly introduced or younger animals. This state of affairs has been noticed in Europe after vaccination against rouget, or swine erysipelas, a disease of swine thus far not found in this country.* While it is not of sufficient importance to militate against the use of successful vaccines, it should nevertheless not be lost sight of in a final estimate of the value of vaccination methods as a whole. Our observations on hog cholera lead me to believe that even if a fairly successful and cheap method of vaccination against hog cholera could be devised the result would be that a number of animals would contract a chronic type of the disease after infection, and these would have all the objectionable features of worthless animals scattering infection about for months.

In all efforts directed towards the production of a certain refractory state of the animal body towards certain disease germs, in other words, in all attempts at preventive inoculation, we must bear in mind that the effort expended is in direct proportion to the gravity of the disease

* See p. 91.

to be fought. This is very well illustrated in the experiments on rabbits and guinea-pigs. Only with great difficulty are these animals made refractory to inoculation with hog-cholera bacilli, while it is comparatively easy to induce a high degree of immunity towards swine-plague bacteria. In all experiments with living cultures or the products of cultures there is a retardation of growth of the vaccinated animal and even a loss of weight, which is the more pronounced the more thorough the preventive treatment. The most successful processes are also the most trying to the animal system. The apparently entirely harmless method of injecting the blood serum of immunized animals suffers from two disadvantages—the immunity is said to be transient unless the treatment is followed by an injection of living cultures, and the difficulties of obtaining the blood serum are considerable. In all methods of preventive inoculation, we must reckon with the cost, which should include the loss in weight and thriftiness sustained by the animals. It is safe to assume that any method which claims to produce immunity towards an infectious disease of a serious character with but little expense or injury of any kind should be looked upon with grave suspicion, for the energy required to bring the animal into the refractory state must on general principles balance or overcome the energy exerted by the disease germs on the unprotected animal organism.

That serious diseases of a slowly progressive character can not be cured or prevented by simple means has been amply demonstrated during the past few years by the history of tuberculin. At first supposed to act as a remarkably rapid cure for human tuberculosis, it is now considered of service only in the earliest stages of this disease. Even here its efficiency is brought out only by prolonged use in minute doses and combined with climatic treatment. It is thus shown that the energy expended in curing early cases is certainly very great, and when applied to animal life wholly out of proportion to the value of the animal, even if a cure could be effected. The same principle is applicable to preventive inoculation, with this difference, that it is very much easier to prevent disease than to cure it. The rapidly fatal diseases known as blood diseases or septicæmias seem to be more easily amenable to preventive inoculation. But even as to these (anthrax, rouget), opinions among authorities are divided, and it has been difficult to obtain any trustworthy data. It has already been stated that a chronic, eventually fatal, form of rouget may appear as a result of the vaccination or as a result of infection after vaccination. Hence, all cases of death among vaccinated animals should be carefully studied before any trustworthy knowledge of the effect of vaccines can be obtained.

Bringing all the facts brought out by the swine investigations together, we would suggest that more attention be paid to the collateral causes which contribute to fatality among swine. These collateral causes necessarily change with the climate, the food, and the manner of keeping such animals, and can be investigated only when the diseases pre-

22275—No. 6——8

vail. More attention should be paid to the effect of food, both in pre-disposing to hog cholera and swine plague and in producing diseases of the digestive organs which simulate hog cholera, but which are probably quite easily prevented. A thorough study should also be made of lung worms and ascarides. If these difficulties can be removed or mitigated, and if the diseases due directly to improper feeding be eliminated by more careful attention to fundamental principles, it is more than probable that the infectious swine diseases will largely disappear. This prediction is based on the fact, experimentally determined and already referred to, that swine naturally possess more or less immunity towards both hog-cholera and swine-plague bacteria, and that with vigorous digestion and sound lungs they may be able to resist the infection which can reach them in decently kept surroundings. We do not intend to deny the possibility that now and then highly virulent races of pathogenic bacteria appear, which are reported to sweep everything before them. These we have still to find. In the field our usual experience has been to find obscure diseases not definitely traceable to any disease germ, or else very chronic cases of hog cholera and swine plague. I would therefore urgently recommend a more extensive study of swine diseases in the field, which should include in its scope all causes of disease. Only by such investigations can we hope to determine the various agencies actually at work in different sections of our country in destroying swine, and to estimate more accurately the losses due to pure infection, to animal parasites, and to improper feeding.

INDEX.

	Page.
Affections of swine due to the bacillus of *Rouget*.	92
Agencies destructive to swine	114
Anaërobic bacteria in the liver of pigs.	104, 106
Attenuation of hog-cholera bacteria in cultures of *Proteus vulgaris*	44
by heat.	42
Bacillus butyricus.	105
choleræ suis	9, 17
a	9, 10
β	10
γ	12
δ	13
ε	13
ζ	13
ζ, associated with swine-plague bacteria in pigs	30, 32, 35
ζ, disease in swine produced by	27
ζ, true hog cholera in swine produced by	15
η, description	16
Bacillus coli communis	12, 19, 21, 22, 23, 24, 25, 39
enteriditis	3, 18, 20, 22, 25
disease produced in man by.	17
effect on experimental animals	19
experiments with	19
Karliuski's observations	18
Bacillus fluorescens liquefaciens.	101
Bacillus found in mare after abortion	17
Bacillus typhi abdominalis.	12
murium, description of	19
destruction of mice with	110
experiments with	20
virulence of	21
Bactericidal power of blood serum of rabbits immune to swine plague.	79
Bang, investigations by	91
Blood serum, immunizing powers of	60, 74
Bronchial gland of pigs, bacteria in	104, 106, 107
Broncho-pneumonia in hog-cholera rabbits	85
Danish hog cholera. (*See* Swinepest.)	
Diphtheria in pigeons.	21
Endocarditis in pigs. (*See Rouget.*)	
Fermentative characters of hog-cholera bacteria.	23
Filtrate of agar suspension of hog-cholera bacteria	55
swine-plague bacteria.	71
Flagella on hog-cholera bacteria.	10
Food, importance of, in raising swine	111
necessary proteids in	115

Page.

Forms of disease produced in rabbits with swine-plague bacteria.............. 82
Gelatin, alkalinity of, as affecting hog-cholera bacilli......................... 11
Hog-cholera bacillus, stability of... 110
    varieties of... 9
Hog-cholera bacilli, dissemination in the organs of pigs 104
    effect on pigeons... 101
    eliminated from organs of immune rabbits............... 62
    immunity produced with sterilized agar cultures......... 54, 55
        blood serum.................. 61, 64, 65
        bouillon cultures............. 46, 50, 53
        defibrinated blood 57, 59
    in spleens of immune guinea-pigs........................ 64
    subcutaneous inoculation into pigs 103
    virulence not increased by rabbit inoculations......... 102
    general characters...................................... 22
    varieties 17, 26, 109
Hog cholera, guinea-pigs rendered immune toward 47, 50
    pigs rendered immune toward 56
    rabbits rendered immune toward 42, 44
    in domestic animals other than swine...................... 110
    Germany 110
    guinea-pigs.............................. 10, 12, 14, 46, 50, 53, 59
    Maryland.. 28
    mice... 10, 12, 14, 15
    modified type... 27
    pigeons... 14, 57, 100
    Virginia 37
Immunity, definition of 77
    processes employed in its production............................. 78
    produced by attenuated bacteria 43
    production of........................ 43, 44, 46, 50, 56, 65, 66, 68, 72, 79
    toward hog cholera does not protect against swine plague and *vice
    versa* 74
Indol reaction... 29
Jensen, investigations of swine erysipelas by................................. 91
Meat poisoning 17, 109
Metchnikoff, description of the bacillus studied by him...................... 60
    verification of Selander's results by 96
Milk, effect of hog-cholera bacteria on 23
Modification of the septicæmic type of swine plague in rabbits.............. 83
Modified disease produced in rabbits with attenuated hog-cholera bacteria... 85
Multiple subcutaneous abscesses. (*See* Subcutaneous abscesses.)
Nonmotile hog-cholera bacillus.. 16
Nesselfieber. (*See* Urticaria.)
Pathogenesis of hog-cholera bacteria........................ 10, 12, 14, 15, 25, 26, 48
Pericarditis in swine-plague rabbits.. 82, 83
Peritonitis in swine-plague rabbits... 76, 82, 83
Pleuritis in swine-plague rabbits 73, 82, 83
Preventive inoculation 91, 112, 113
Processes employed in the production of immunity........................... 78
Proteus vulgaris 25, 44
Pseudo-tuberculosis, bacilli of... 93
    in rabbits and guinea-pigs............................... 93
    produced by hog-cholera bacteria in guinea-pigs.. 48, 50, 53, 87
    in rabbits... 88

Page.

Rabbits, disease produced by hog-cholera bacteria in........................ 10
Rabbit septicæmia, bacillus of 75, 102
          experiments to increase virulence of 102
Rouget, endocarditis accompanying.. 91
      varieties of disease caused by the bacillus of 92
      vaccination disease.. 91
Selander's experiments to increase virulence of hog-cholera bacteria.......... 98
          with the blood of hog-cholera pigeons 57
Septicæmia amenable to preventive inoculation 113
      in swine-plague rabbits... 82
Subcutaneous abscesses in swine-plague rabbits...................... 67, 70, 73, 84
Subperitoneal abscesses in swine-plague rabbits....................... 67, 84, 97
Swine, collateral causes of disease among.................................... 113
      not very susceptible to hog cholera or swine plague 111
Swine erysipelas... 91
Swine pest.. 96
Swine plague, disease produced in partially immune rabbits 66, 69, 73, 83, 84
      guinea-pigs immune to 73, 76
      methods of producing immunity towards................... 65, 68, 72
      rabbits immune to 66, 68, 72, 76
      varieties of disease produced in rabbits........................ 82
Swine-plague bacteria 30, 32, 35, 65
      at seat of inoculation in pigs........................... 106
      subcutaneous inoculation into pigs.................... 106
Toxic effect of blood of hog-cholera guinea-pigs............................. 60
      bouillon cultures of hog-cholera bacteria.................. 47, 51, 52
      defibrinated blood of swine-plague rabbits. 72
      filtrate of hog-cholera and swine-plague cultures............. 55, 72
      sterilized agar cultures of hog-cholera bacteria............ 53, 54, 55
      swine-plague bacteria 65, 69
Tubercle-like bodies in hog-cholera guinea-pigs. (*See* Pseudo-tuberculosis.)
Tuberculin 113
Type of disease, cause of variation in 89
Ulcerated intestine of rabbits affected with hog-cholera 85
Urticaria in swine in Denmark.. 92
      Germany 92
Variability of swine plague and hog cholera.................................. 81
Vaccination of rabbits with artificially attenuated hog-cholera bacteria 42
      bacillus ζ..................................... 43
      dangers of.. 71
Virulence of hog-cholera bacteria 97
      experiments to increase................... 99

O

www.ingramcontent.com/pod-product-compliance
Lightning Source LLC
Chambersburg PA
CBHW021823190326
41518CB00007B/713